Arterial Blood Gas Interpretation for the ACEM Fellowship Exam:
25 worked examples

Dr Luke D. Lawton

BAppSc (Biochemistry), MBBS (Hons) MPH (Aeromedical Retrieval) FACEM
Staff Specialist Emergency Medicine, Redcliffe Hospital
Retrieval Consultant, Careflight Medical Services
Clinical Lecturer, the University of Queensland
Queensland, Australia

Dr Corinne A. Ryan

BAppSc (Biochemistry) MBBS (Hons) FRACP
Oncology Fellow, Redcliffe Hospital
Clinical Lecturer, the University of Queensland
Queensland, Australia

CALL the Doctors Publishing
80/20 Donkin St
Brisbane, Queensland, Australia

© 2013 CALL the Doctors Publishing

National Library of Australia Cataloguing-in-Publication Data

Lawton, Luke Douglas.
Arterial Blood Gas Interpretation for the ACEM Fellowship Exam: 25 worked examples/
Luke D Lawton, Corinne A Ryan
1st Ed
ISBN: 978 0 9924245 0 3 (pbk)

Biochemistry interpretation
Emergency Medicine
Arterial Blood Gas

Publisher: Luke Lawton
Publishing Services Manager: Corinne Ryan
Editor in Chief: Corinne Ryan
Cover Design: Luke Lawton

Contents

Foreword: how to use this book

"I'm not telling you it's going to be easy – I'm telling you it's going to be worth it."

Arthur T Williams

This book consists of 25 original questions in the style of the VAQ paper from the ACEM Fellowship exam. The questions are divided by complexity to allow each reader to judge their progress.

Each question has a worked answer, done to time (10 minutes), and some brief comments either concerning the subject matter or specific notes about the content or structure of the suggested answer. These questions have been written with specific regard to ACEM exam terminology, which can be found in the Examinations Handbook and is reproduced in the appendices here.

The worked answers are suggestions about how to produce an above average answer. Some candidates may have different opinions about the correct content of an answer. This is the nature of emergency medicine. Some of the ACEM questions are quite nebulous in their requests, particularly when it comes to "interpretation". However, all the questions given illustrate a reproducible structure for answering in the real exam, as well as practical application of relevant calculations. If you are beginning your exam preparation and are lost as to how to proceed with an ABG question the answers within should provide some guidance as to what is expected. If you are nearing the exam and within ten minutes can produce something approximating the answers in this book then you should be confident of performing well on the big day.

If you are not sitting the ACEM exam but are looking for guidance or practice on answering blood gas questions (perhaps as either as a medical student an MCEM candidate) this book still provides a useful and practical structure for interpreting the Arterial Blood Gas. Annotations of kPa equivalents for mmHg accompany all questions for those practicing in the British Isles.

We should stress that this book does not aim nor purport to offer clinical advice as to the treatment of patients. The references given for each particular section are not necessarily definitive but have been chosen because they illustrate the principles described and provide a useful springboard for further study.

The inspiration to write this book was the complete lack of any formal guides, worked examples or concentrated source of arterial blood gas questions when it came to preparing for the exam. Experience shows that these questions are a source of confusion and anxiety for candidates. This book is intended to illustrate a simple, reproducible method for answering well. We hope you find it useful.

Luke Lawton and Corinne Ryan

Abbreviations

A-a	Alveolar-arterial (oxygen gradient)
ABG	arterial blood gas
ACEM	Australasian College for Emergency Medicine
AF	atrial fibrillation
AG	anion gap
APO	acute pulmonary oedema
ARDS	adult respiratory distress syndrome
Ax	assessment
AXR	abdominal x-ray
BiPAP	bilevel positive airway pressure
BSL	blood sugar level
Ca	cancer
COPD	chronic obstructive pulmonary disease
CPAP	continuous positive airway pressure
CT	computed tomography (scan)
CTPA	computed tomography pulmonary angiogram
CVA	cerebrovascular accident
CVC	central venous catheter
CXR	chest x-ray

(D)Dx	(differential) diagnosis
DKA	diabetic ketoacidosis
DR	delta ratio
DVT	deep vein thrombosis
ECG	electrocardiogram
Ex	examination (physical)
FACEM	Fellow of the Australasian College for Emergency Medicine
FAST	focused abdominal sonography in trauma
FRACP	Fellow of the Royal Australian College of Physicians
GCS	Glasgow coma score
Hx	history
ICH	intracerebral haemorrhage
IPPV	invasive positive airway pressure
Ix	investigation
LRTI	lower respiratory tract infection
N	normal
NAGMA	normal (or non) anion gap metabolic acidosis
NAI	non-accidental injury
PE	pulmonary embolus
RAGMA	raised anion gap metabolic acidosis
RTA	renal tubular acidosis

SAH	subarachnoid haemorrhage
SIRS	systemic inflammatory response syndrome
VAQ	visual aid questions
VBG	venous blood gas
V/Q	ventilation perfusion ratio

A general approach to arterial blood gas questions

"Simplicity is the ultimate sophistication"
Leonardo Da Vinci

Candidates become flustered on the blood gas analysis questions administered as part of the ACEM Fellowship exam. However ABG questions offer an excellent opportunity to score heavily in the VAQ section if approached in a systematic fashion. They are all quite similar, without the subjective vagaries of the "describe and interpret a picture" type questions and this lends itself to a systematic, repeatable approach. Ideally a candidate should write essentially the same answer every time, and just "plug in" the numbers from the question. Many examinees get flustered by ABG questions because of a lack of understanding about the questions themselves, or by lacking the discipline to develop a comprehensive, simple, reproducible approach to the questions in their preparation.

There are two issues to think about when preparing for these questions.

Why is there an ABG question on the VAQ paper?

This is not rhetoric.

The primary reason ABG questions appear is because the question *illustrates a disturbance of acid-base metabolism that the examiners would like you to identify and describe*. There are subsequently two conclusions to draw:

i) There *will be* an acid base disturbance present in the question
ii) To pass, you need to describe it. Everything else is window dressing for bonus marks.

Within this context then, there are very few things you need to be able to talk about to answer the question properly:

i)   metabolic acidosis
        • raised anion gap
        • normal anion gap
ii)   metabolic alkalosis
iii)   respiratory acidosis
iv)   respiratory alkalosis

The list really *is* that small: 4 disorders. Having a reproducible and customizable differential for each is the key to answering the ABG question well. Such lists abound on the internet and are easily customisable. A list provided by one of the authors as an example can be found in appendix 5.

A second thought to consider about the ABG question pertains to ease of marking. In general there are a large number of calculations which can be performed by candidates and therefore marked objectively by examiners. This is an advantage for both. For examiners the question is standardized and easy to mark. For candidates the question is standardized and easy to milk. Performing calculations appropriately is critical to achieving a high mark on ABG questions. Common calculations are:

i)   anion gap
ii)   delta ratio
iii)   A-a gradient
iv)   osmolar gap
v)   corrected CO2/HCO3- values
vi)   electrolyte corrections, particularly Na and K
vii)   urea:creatinine ratios

You should anticipate that examiners will be looking for these calculations with pen poised as they read your answer. However it is important not to get too obsessive about decimal points for the calculations. The exam is conducted under extreme time pressure without a calculator. Some approximation is acceptable and will be evident in the worked examples which follow. Accuracy is important, precision less so.

How do I answer to maximize my mark?

Begin by addressing the central acid/base imbalance

Put the question into context by ensuring you read the stem. This often gives you the diagnosis and simultaneously sets up your interpretation, yet most candidates ignore it except as the herald of a difficult ABG question. Two examples from real exams follow with comments:

Example 1.

"A 59 year old <u>obese</u> man receives 5 mg of <u>intravenous morphine</u> for analgesia for abdominal pain. Thirty minutes later, <u>his GCS has fallen to 12</u> and investigations are performed." (VAQ 2009.2)

Straight away before even looking at the numbers candidates should note the following cues:
- i)   *Obese – likely sleep apnoea, probably chronic CO_2 retention*
- ii)   *IV morphine and decreased GCS causing acute hypoventilation and acute on chronic respiratory acidosis. This creates a need to correct HCO_3^- for both acute and chronic CO_2 retention.*
- iii) *This is an iatrogenic insult. Both reversal and quality assurance should be mentioned in the interpretation*

Unsurprisingly when worked these observations are exactly what the blood gas reveals.

Example 2.

"A 45 year old man with <u>type 1 diabetes mellitus</u> is brought in by ambulance with an <u>altered conscious state</u>" (VAQ 2010.1)

Straight away candidates can infer:
- i)   *A diagnosis of DKA, and expect a showing RAGMA, Delta Ratio ~1, and profound hyperglycaemia*
- ii)   *The possibility of a concurrent lactic acidosis due to shock/sepsis*
- iii) *Probable hyponatraemia. Na^+ should be corrected for glucose*
- iv)   *The need for resuscitation and insulin infusion and the risk of cerebral oedema should all be mentioned in the interpretation.*

Again these are all elements of the correct answer on working the question and reviewing the college's examination report.

Put on some window dressing as you work

Comment on minor issues as the question progresses. Don't leave everything to the end because doing so becomes rapidly overwhelming. If your comments on minor issues are inserted in the question as you work, you pick up the marks there and then and are free to concentrate on the major acid base disturbance in your interpretation. Examples of minor issues include:

i)   appropriate compensation for the major disorder
ii)   A-a gradient
iii)   electrolytes:
- correct Na^+ (for glucose)
- correct K^+ (for pH)
- comment on the Cl^- if it's low. My standard comment is *"electrical equilibration for RAGMA"*
- calculate Urea:Creatinine ratio. Usually it suggests pre-renal failure
- calculated osmolality

Further information may be obtained in appendices 2 and 3.

Work through in a systematic and stepwise fashion that is easy for the examiners to follow.

Think for a moment: the examiners have to mark 50+ answers each exam. If your answer is illegible, written in hieroglyphics or algebra, or generally a dog's breakfast you are likely to be the victim of a negative free-floating marking bias. Speed is at a premium, but make sure that any abbreviations you use are common medical lexicon. A good place to start for what may be considered acceptable abbreviations is the abbreviation list contained in this book.

Essentially if you make it harder for the examiners they are likely to mark you accordingly. (If you need to be told to write legibly you should not be sitting this exam.) See the following section for a suggested structure to maximise your answer's chances.

Show perspective.

Use adjectives. pH of 7.26 and 6.9 are both "acidaemias". One is a "mild" acidaemia, one is "life threatening". You want to be a consultant. Prove it.

Show sophistication: **MUDPILES** *is for* **MEDICAL STUDENTS**

Have a sophisticated interpretation that conveys understanding, not wrote learning. MUDPILES (causes of RAGMA) almost all present as a lactic acidosis. Writing "MUDPILES" as the differential for a RAGMA is as intellectual as withdrawing money from the "ATM Machine".

There are only four significant causes of a RAGMA[1]:

    i)   ketoacids
                    i.   starvation
                    ii. diabetes
                    iii. profound dehydration (hyperemesis gravidarum, children)

    ii)   lactic acid[2]
                    i.   type A: tissue hypoperfusion (shock, anaemia, haemorrhage, SMA occlusion etc)
                    ii. type B_1: liver failure, sepsis
                    iii. type B_2: drugs (all the MUDPILES ones)
                    iv. type B_3: inborn errors of metabolism. (*Editor's note: thankfully it's not a PICU fellowship exam…*)

    iii)   uraemic renal failure

Rather than just reiterate a list, put something on your paper that shows understanding. Consultant opinions are characterized by clarity and sophistication.

---

[1] Kraut JA and Madias NE. Serum nion gap: its uses and limitations in clinical medicine. *Clin J Am Soc Neprhol* 2007; 2:162-174

[2] Kriesberg RA. Lactate homeostasis and lactic acidosis. *Ann Intern Med* 1980; 92: 227-37

Specific structure: actually writing the answer

"Success isn't always about 'greatness'. It's about consistency".
Dwayne "The Rock" Johnson

In the worked examples section of this book the following format is used to answer ABG questions. It is adapted and altered as appropriate for each individual question, but the same process is followed for all.

All calculations are included *in vitro* and the answer progresses logically. Numeric headers guide the examiners through the answer. Specific notes on several of the calculations are provided in the appendices.

1. **Acid base balance**
   - Comment on the pH: *acidaemia, alkalaemia*
   - Comment on the CO_2: *hyper/hypocarbia, respiratory acidosis/alkalosis*
   - Comment on the bicarbonate: *high/low, metabolic acidosis/alkalosis*
   - Decide on the major disturbance. *This always is in the same direction as the pH.*
   - Correct the dependent variable for the major disturbance. *If the primary disturbance is metabolic, correct the CO_2 for the HCO_3^-. If the primary disturbance is respiratory, correct the HCO_3^- for the CO_2. If both disturbances occur in the same direction, correct each for each other to demonstrate independence.*
   - If there is a metabolic acidosis calculate an anion gap and delta ratio
   - Give a summary statement of the above, for example: *"Isolated RAGMA, appropriate respiratory compensation"*

2. Oxygenation
- Calculate A-a gradient
- Make a comment. This can be standardized across answers, unless it is crucial to the acid-base disturbance (in the case of a respiratory acidosis and type I or type II respiratory failure)

3. Electrolytes and other numbers:
- Comment on all minor abnormalities
- Don't forget:
    - i.   Correct Na^+ for glucose
    - ii.   Correct K^+ for pH
    - iii. Calculate U:C ratio
    - iv.   Calculate osmolality and osmolar gap

4. Interpret and synthesise

- This is where the major list for the acid base disturbance comes into play. Highlight the major disturbances in context.

"Interpret" can be a pretty nebulous term. The college has its own definition but the real key is to relate your answer back to the stem. To do this, ask:

What is the diagnosis/differential?

- write the appropriate list and identify the most likely diagnosis

What other tests do you need to sort out the differential diagnosis?

- glucose for RAGMA if not given (indicates possible DKA)
- CT brain for obtundation
- Septic workup if sepsis is suspected
- Drug levels if the patient has taken an overdose (don't forget paracetamol!)
- Essentially ask yourself "what would I want the intern to have ordered for this patient?"

What are the implications of your findings?

- Likely prognosis
- Need for emergent management: antidotes, O_2, intubation, intensive care and other niceties

If you practice hard and prepare well all of that should take you a little under 10 minutes.

Good luck!

THE EASY PROBLEMS

"'Excellent!' I cried. 'Elementary,' said he."
Sir Arthur Conan Doyle

PROBLEM 1.

A 6 year old boy presents with his parents with a five day history of profound watery diarrhoea. He is confused and drowsy at triage with a capillary refill time of 4 seconds. He is transferred to the resuscitation area and a venous blood gas is immediately taken.

Describe and interpret the results.   (100%)

pH	7.22		
PO_2	45	mmHg (5.99	kPa)
PCO_2	18	mmHg (2.4	kPa)
HCO_3^-	6	mmol/L	
BE	-18		
Na^+	140	mmol/L	
K^+	4.0	mmol/L	
Cl^-	120	mmol/L	
Glucose	2.6	mmol/L	
Urea	13	mmol/L	
Creatinine	125	μmol/L	
Lactate	1.5	mmol/L	

2

ANSWER

1. Acid Base Balance

Moderate acidaemia
Profound hypocapnoea
       Respiratory alkalosis

Critically low bicarbonate
       Metabolic acidosis
       Expect CO_2 = 8 + 1.5 x 6
                         = 17
       .: appropriate compensation

       AG   = 140 − 120 − 6 = 14
       DR   = (14 − 12)/(24-6)
                = 2/18
                <0.4 therefore Dx NAGMA

Therefore diagnose NAGMA with appropriate respiratory compensation

2. Oxygenation

A-a gradient incalculable: venous BG

3. Electrolytes

Normonatraemia
Normokalaemia
       Expect K^+ pH 7.22   = 5 + 2 x 0.5
                                  = 6mmol/L
       Relative hypokalaemia, watch K^+ as pH corrected
Normochloraemia
Elevated renal indices
       U:C > 100 suggests pre-renal failure
Critically low glucose
       → requires urgent correction with 5ml/kg 10% dextrose

4. Interpret

6yo male, approx weight 20kg
Major disturbance is NAGMA
        Appropriate resp compensation
        Causes of NAGMA to consider in this child
               HCO_3^- loss ***most likely diagnosis***

Diarrhoea

RTA (elevated renal indices)

Endocrinopathy

Addison's less likely given Na/K normal

Fistula

Pancreaticoduodenal

Uretoenteric

Drug overdose

Acetazolamide/spironolactone

Check parent's meds

Be aware NAI/neglect

delayed presentation

other causes of drowsiness (eg head injury)

Will need admission, fluid resuscitation 10-20ml/kg 0.9% NaCl as per APLS

Watch electrolytes

Seek and treat cause

If persistent diarrhoea consider stool culture

COMMENTS

A 6 year old boy presents with his parents with a five day history of profound watery diarrhoea. He is confused and drowsy at triage with a capillary refill time of 4 seconds. He is transferred to the resuscitation area and a venous blood gas is immediately taken.

Attention should be paid to the relevant indicators given in the stem of the question. In generally diarrhoea is associated with HCO_3^- loss and a resultant NAGMA[3]. There are also clear indicators that this child is shocked and unwell.

As may also be expected from the stem there is biochemical evidence of dehydration. Note has been made of this in the main body of the answer so as to be able to focus on the major issues in the interpretation phase. Likewise the critically low glucose has been noted and addressed, thus securing these marks prior to discussing the main metabolic disturbance. Such a critical issue is likely to represent pass/fail criteria in the exam and should not be omitted.

Finally, as this is a paediatric patient there are several "niceties" that should be observed and noted somewhere. A weight is essential (remembering the formula of (2 x age in years + 8)[4]. For a seriously unwell child to wait five days before presenting to hospital would be unusual, and raises the spectre of a child at risk. This should also be mentioned by strong candidates.

---

[3] Perez GO, Oster JR and Rogers A. Acid-base disturbances in gastrointestinal disease. *Dig Dis Sci* 1978; 32(9): 1033-43

[4] Luten R and Zartisky A. The sophistication of simplicity....optimizing emergency dosing. *Acad Emerg Med* 2008; 15(5): 461 - 5

PROBLEM 2

A 26 year old woman with type 1 diabetes is brought into the emergency room by the ambulance with a decreased level of consciousness. Her pathology is shown below.

Describe and interpret the results.   (100%)

Serum ABG and biochemistry

FiO_2	0.5	
pH	7.05	
PO_2	186	mmHg (24.8k kPa)
PCO_2	66	mmHg (8.79 kPa)
HCO_3^-	16	mmol/L
BE	-8	
Sats	99%	
Na^+	132	mmol/L
K^+	5.0	mmol/L
Cl^-	92	mmol/L
Urea	15	mmol/L
Creatinine	227	μmol/L
Glucose	50.9	mmol/L

ANSWER

1. Acid base status

Critical acidaemia (pH 7.05)
Moderate hypercapnoea
       Respiratory Acidosis
       Expect HCO_3^- to be ~24-26 acutely for compensation

Moderately low bicarbonate and negative base excess
       Metabolic acidosis
       Expect pCO_2 = (20 + 0.8x18)
                    = 35mmHg
       Anion Gap   = 132 – (92+16)
                    = 24
       Delta Ratio   = (24 – 12)/(24-16)
                    = 12/8 =1.5 (<2.0)

               Therefore Dx isolated RAGMA

Overall picture: combined respiratory acidosis and RAGMA

2. Oxygenation

A-a gradient for FiO_2 = 50%
            = 350 – (1.25x 66) – 186
            = 350 - ~80 – 186
            = 270 – 186
            = 64
Expect A-a grad for 45yo man = 45/4 + 4 ie 15mmHg

Therefore elevated A-a gradient suggesting respiratory shunt
       ARDS (SIRS/aspiration)
       Interstitial lung disease
       Other respiratory membrane disease (eg LRTI/APO)
       PE

3. Biochemistry

Normonatraemia
       correct for glucose Na^+ = 131 + 45/3 = 146
       mild elevation ?dehydrational
Mild hyperkalaemia

7

expect K$^+$ for ph 7.05 = 5 + 3x0.5 = 6.5 mmol/L
therefore relatively hypokalaemic and need to watch K$^+$ as
resuscitated
Mild hyperchloraemia
Electrical equilibration RAGMA
Elevated renal indices
U:C ratio >50%
Likely some pre-renal failure 2ndary to decreased perfusion
(dehydration/sepsis)
May also have element of intrinsic renal failure (diabetic
nephropathy)

Profound hyperglycaemia
Osmolality$_c$ = 2 x 131 + 50 + 15 = 262 + 65
= 327mmol/Kg
HIGH
risk of cerebral oedema

4. Interpretation

This pt is critically ill:
Dx RAGMA + hyperglycaemia
Critical DKA
Needs ICU, insulin infusion and fluid/K replacement
Other causes RAGMA to consider here
Renal failure (present)
Increased lactate: check level
A – hypoperfusion, secondary to
hypovolaemia
B$_1$ – disease states (hepatic failure, sepsis)
B$_2$ – drugs especially metformin or other
OD

Elevated A-a gradient and increased CO_2
Respiratory insufficiency
Concern RE aspiration given clinical history and decreased
LOC
Check CXR
Septic screen as indicated

Decreased LOC and hyperosmolality
Concern RE cerebral oedema

A-a gradient in kPa

A-a grad
$$= 50\text{kPa} - 1.25 \times 8.79 - 24.8$$
$$= 50 - 11 - 25$$
$$= 14 \text{ kPa}$$

Elevated

COMMENTS

A 26 year old man with type 1 diabetes is brought into the emergency room by the ambulance with a decreased level of consciousness. Her pathology is shown below.

Even a brief review of the stem will reveal the likely answer – diabetic ketoacidosis. Systemic working of the question reveals a RAGMA and a respiratory acidosis with a V/Q mismatch. To demonstrate the dual disturbance appropriate calculations are performed for both respiratory and metabolic compensation. The resultant inequity in CO_2 and HCO_3^- clearly reveals two independent processes.

This question is an illustration of several principles of answering examination ABG questions. Every calculation that can be performed should be, including in this case a calculated serum osmolality. The importance of serum osmolality in the generation of cerebral oedema remains controversial[5,6], but strong candidates will perform the calculation and hopefully pick up a mark for doing so. Corrections of Na^+ for glucose and K^+ for pH should likewise be performed. Other differentials for the RAGMA should also be mentioned, in particular lactate, which has been long described as a contributor to the acidosis in some cases of DKA[7]. As part of the interpretation the need to check a lactate therefore requires articulation. In the context of a raised A-a gradient a chest x-ray is indicated, and strong candidates will interpret the major findings as an indication for insulin therapy and disposition to a critical care area.

---

[5] Tiwari LK, Jayashree M, Singhi S. Risk factors for cerebral edema in diabetic ketoacidosis in a developing country: role of fluid refractory shock. *Paed Crit Care Med* 2012; 13(2):e91-6

[6] Hoorn EJ and Zietse R. Cerebral oedema in adult diabetic ketoacidosis: the importance of effective serum osmolality. *Netherlands J Med* 2010; 68(12):439

[7] Watkins PJ, Smith JS, Fitzgerald MG and Malins JM. Lactic Acidosis in Diabetes. *BMJ* 1969; 1(5646): 744 - 747

PROBLEM 3

A 22 year old girl presents to ED complaining of chest pain, parasthesiae and dizziness. She has no medical history of note.

Her vital signs are:

HR	105	/min
BP	120/80	mmHg
RR	26	/min
Sats	100%	RA
Temp	37	°C

An arterial blood gas is taken by your intern prior to registrar review and is shown below.

Describe and interpret the results.   (100%)

FiO_2	0.21		
pH	7.49		
PO_2	107	mmHg (14.26	kPa)
PCO_2	28	mmHg (3.7	kPa)
HCO_3^-	22	mmol/L	
BE	+1		
O_2 Sats	100%		
Na^+	135	mmol/L	
K^+	4.6	mmol/L	
Cl^-	99	mmol/L	
Glucose	5.2	mmol/L	
Lactate	0.72	mmol/L	

11

ANSWER

1. Acid base status

Mild alkalaemia (pH 7.49)

Mild-moderate hypocapnoea
  Respiratory alkalosis
  Expect HCO_3^- for CO_2 28
      Acute $= 24 - 1 \times 2 = 22mmHg$
      Chronic $= 24 - 1 \times 5 = 19mmHg$

Mild decrease in HCO_3^-
  Metabolic acidosis
  Consistent with acute compensation for respiratory alkalosis

Therefore diagnose:
  acute respiratory alkalosis
  appropriate respiratory compensation

2. Oxygenation

Mild hyperoxaemia on RA
  Secondary to hyperventilation

A-a gradient $= 150 - 1.25 \times 28 - 107$
        $= 150 - 35 - 107$
        $= 150 - 42$
        $= 8$

Expect A-a gradient 22yo female $= 22/4 + 4 = 5.5 + 4 = 9.5$
Not elevated
No V/Q mismatch exists

3. Electrolytes

Normonatraemia
Mild hyperkalaemia
    Expect K^+ for pH $= 5 - 1 \times 0.5 = 4.5$ mmol/L
    K^+ level appropriate for pH
Trivial hypochloraemia
Normoglycaemia
Normal lactate

4. Interpret

ABG shows
>acute primary respiratory alkalosis with hyperoxaemia
>appropriate metabolic compensation
>no V/Q mismatch and normal vitals
>>serious pathology very unlikely

Major diagnosis is anxiety/panic attack
However multiple differentials possible
>PE: assess risk carefully PERC/Well's and Ix as appropriate
>Pneumothorax: CXR
>Drug effect or withdrawal state
>>Alcohol
>>MDMA
>>Cocaine
>>$\rightarrow$ clarify history and evaluate for toxidrome
>Pericarditis
>>Check ECG
>Pancreatitis
>>Check lipase

Issue of quality assurance
>Indication for ABG (invasive, painful) by junior doctor
>Investigate and educate as appropriate

A-a gradient in kPa

A-a grad
>$= 21\text{kPa} - 1.25 \times 3.7 - 14.26$
>$= 21 - 4.6 - 14.3$
>$= 21 - 18.9$
>$= 2.1$

Normal (<2.5kPa)

13

COMMENT

A 22 year old girl presents to ED complaining of chest pain, parasthesiae and dizziness. She has no medical history of note.

An arterial blood gas is taken by your intern prior to registrar review and is shown below. Describe and interpret the results.

This is a slightly unusual stem. The first part suggests a very common ED presentation: that of acute anxiety. Consultant level perspective is given by showing awareness that the normal vital signs make occult pathology very unlikely in a young patient. However a number of serious differentials do exist for this presentation and should be mentioned in conjunction with the appropriate calculations being performed. This is essential in achieving a higher mark, and a key here is delineating extra tests which may be indicated, such as an ECG, chest x-ray or a lipase level.

The other issue which the stem prompts candidates to mention is one of quality assurance. Such issues occur intermittently within the exam, but the chance to comment for extra marks should never be missed. In this case performance of an arterial stab for blood gas analysis would seem to be overzealous, and this is an issue which should be followed up.

14

PROBLEM 4

An 87 year old man presents with severe central abdominal pain, vomiting and hypertension for 6 hours. He has a past history of paroxysmal atrial fibrillation and ischaemic heart disease.

His vital signs are:

HR	110	/min
BP	98/68	mmHg
RR	36	/min
T	36.5	°C
Sats	100%	28% O_2

Arterial blood gas analysis is shown below.

Describe and interpret the results.   (100%)

FiO_2	0.28		
pH	7.20		
PO_2	98	mmHg (13.0	kPa)
PCO_2	30	mmHg (4.0	kPa)
HCO_3^-	14	mmol/L	
BE	-10		
O_2 Sats	99%		
Na^+	142	mmol/L	
K^+	5.5	mmol/L	
Cl^-	106	mmol/L	
Lactate	6.4	mmol/L	

15

ANSWER

1. Acid base status

Moderate acidaemia (pH 7.20)
Mild hypocapnoea
        Dx respiratory alkalosis

Moderately low bicarbonate
        Significant metabolic acidosis
                expect CO_2 $= 8 + 1.5 \times 14$
                               $= 29$ mmHg
            .: appropriate respiratory compensation

    AG $= 142 - 14 - 106$
                     $= 22 \rightarrow$ RAGMA
    delta ratio $= (22\text{-}12)/(24\text{-}14)$
                     $= 10/10$
                     $= 1.0$.: Dx isolated RAGMA
                     Fits with BE -10 for HCO_0^- 14

Therefore diagnose RAGMA and compensatory respiratory alkalosis.

2. Oxygenation

$pO_2 \rightarrow$ normoxia

A-a gradient $= 225 - 1.25 \times 30 - 98$
                  $= 225 - 37 - 98$
                  $= 225 - 135$
                  $= 100$
Expect for 87yo male $87/4 + 4 = 22 + 4 = 26$
        .: Dx elevated A-a gradient $\rightarrow$ Dx V/Q mismatch
                ARDS (SIRS/aspiration)
                Interstitial lung disease
                Other respiratory membrane disease (eg LRTI/APO)
                PE

3. Electrolytes

Normonatraemia
Mild hyperkalaemia
        for pH 7.20 expect $5.0 + 2 \times 0.5 = 6.0$mmol/L
                K^+ mildly low for pH

16

watch K^+ as pH corrected and correct if needed
Normochloraemia
Significant hyperlactataemia
> Likely cause of RAGMA:
>> Type A - tissue hypoxia/shock
>>> Note hypotension (bleeding, dehydration)
>>> Vasculopath → possible mesenteric ischaemia
>> Type B_1 sepsis/underlying disease (eg liver failure)
>> Type B_2 drug OD
>>> Possible drug toxicity: metformin, isoniazid, salycilates, paraldehyde, toxic alcohols etc

4. Interpret

ABG shows:
> RAGMA with respiratory compensation
>> Hyperlactataemia (lactic acidosis)
> Mild relative hypokalaemia
> V/Q mismatch
>> Given Hx vomiting concern RE aspiration/ARDS

87yo male, vasculopath, AF
> Concern RE catastrophic Dx
>> Ischaemic mesentery
>> Leaking AAA
>> Bowel obstruction
>> Perforated viscus
>> Severe sepsis
> Guarded prognosis
>> Needs further evaluation
>>> Renal indices/LFTs
>>> Septic workup
>> Consult surgeons and radiology
>>> CT – AAA, pneumatosis, free air
>>> Bedside U/SS if available
>> Check for advanced health directive RE limits of care

A-a gradient in kPa

A-a grad
$$= 28kPa - 1.25 \times 4 - 13$$
$$= 28 - 5 - 13$$
$$= 28 - 18$$
$$= 10$$

Elevated (>4.0 kPa)

COMMENTS

An 87 year old man presents with severe central abdominal pain, vomiting and hypertension for 6 hours. He has a past history of paroxysmal atrial fibrillation and ischaemic heart disease.

In this straightforward question the stem highlights a potentially catastrophic clinical situation. The patient is old, possesses several vascular risk factors and has severe abdominal pain. Candidates who cannot generate an appropriate differential list for this situation should not be sitting the exam! A severe metabolic acidosis should be expected as a complication of most of these illnesses[8] and this is indeed what the working reveals. Further investigations may be required to evaluate and these are briefly discussed.

The V/Q mismatch is dealt with "in question", highlighting the possibility of an aspiration, and again the K^+ is corrected appropriately. Perspective is shown by acknowledging the patient's age and guarded prognosis, and a statement that aggressive intervention may be unwarranted.

---

[8] Lange H and Jackel R. Usefulness of plasma lactate concentration in the diagnosis of acute abdominal disease. *Eur J Surg* 1994; 160(6-7): 381-84

PROBLEM 5

Your retrieval team is tasked to pick up an 80 year old male with COPD and pulmonary oedema from a rural hospital because he requires non invasive ventilation. He has been aggressively treated with loop diuretics and nebulized salbutamol.

His venous blood gas on your arrival is shown below.

Describe and interpret the results.   (100%)

FiO_2	0.21		
pH	7.58		
PO_2	66	mmHg (8.8	kPa)
PCO_2	48	mmHg (6.4	kPa)
HCO_3^-	38	mmol/L	
BE	+14		
O_2 Sats	89%		
Na^+	126	mmol/L	
K^+	2.2	mmol/L	
Cl^-	82	mmol/L	
Urea	8.7	mmol/L	
Creatinine	142	μmol/L	

ANSWER

1. Acid base status

Severe alkalaemia (pH 7.58)

Mild hypercapnoea
> Respiratory acidosis

Significantly elevated bicarbonate
> Metabolic alkalosis
>> expect CO_2 = 20 + 0.7 x 38
>> = 20 + 28
>> = 48 mmHg
> therefore appropriate respiratory compensation
> AG/DR N/A for alkalosis

Therefore diagnose primary metabolic alkalosis with appropriate respiratory compensation

2. Oxygenation

Cannot be interpreted – venous sample

3. Electrolytes

Moderate hyponatraemia
> Renal excretion secondary diuretics

Severe hypokalaemia
> for pH 7.58 expect K^+ = 5.0 - 2 x 0.5 = 4.0mmol/L
>> .: K^+ still critically low when corrected for pH
>> Likely due to diuretic use
>> High risk arrhythmias/respiratory failure
>> Needs immediate correction with high dose IV K^+

Severe hypochloraemia
> Renal excretion due to diuretics

Renal indices elevated
> Urea: Creatinine ratio ~50
> Implies element of intrinsic renal failure. May be medication induced

21

4. Interpret

ABG shows:
> Primary metabolic alkalosis with compensatory respiratory acidosis
> Multiple electrolyte deficiencies
> Life threatening hypokalaemia
> Renal failure

Metabolic alkalosis
> Diuretic induced (frusemide)
> Also consider
>> Endocrinopathies (Cushing's, Bartter's Sx)
>> HCl loss – vomiting

Respiratory acidosis
> Compensatory
> No acute indication for BiPAP at this stage: may worsen alkalaemia
> Needs diuretics ceased
> Urgent K+ replacement, telemetry

Iatrogenic illness
> Feed back to treating doctor
> Quality assurance procedures

COMMENTS

Your retrieval team is tasked to pick up an <u>80 year old male</u> with <u>COPD and pulmonary oedema</u> from a rural hospital because he <u>requires non invasive ventilation</u>. He has been aggressively treated with <u>loop diuretics and nebulized salbutamol</u>.

This question is a little unusual in that the primary derangement discovered is actually an alkalaemia rather than the expected acidaemia. There is a respiratory acidosis, but this is *against* the primary direction of the pH change. Therefore it is compensatory, rather than primary. In this case the benefit of bi-level positive airway pressure is extremely dubious[9], as it will worsen the alkalaemia. There has clearly been a misinterpretation of the patient's pathology by the treating clinician.

The question also reveals multiple electrolyte elements, and an element of intrinsic renal failure[10]. There are many differentials which can be mentioned but again this question highlights the importance of returning to the stem when interpreting. In this case it highlights the cause of the problem: overzealous administration of a loop diuretic.

Once again exceptional candidates will take the opportunity to mention the issue of quality assurance. This is an iatrogenic pathology and as such should be fed back to the treating clinician.

---

[9] Celikel T, Sungur M, Ceyhan B and Karakurt S. Comparison of noninvasive positive pressure ventilation with standard medical therapy in hypercapnic acute respiratory failure. *Chest* 1998; 114(6): 1636-42

[10] Miller TR, Anderson RJ, Linas SL, Henrich QL, Berns AS, Gabow PA et al. Urinary diagnostic indices in acute renal failure: a prospective study. *Ann Intern Med* 1978; 89(1):47-50

PROBLEM 6

A 28 year old man presents with lethargy and dizziness. He has no medical history of note, and is a regular attendee of the local gym. He reports trying to lose weight over the last month, but denies any other symptoms.

His arterial blood gas is shown below.

Describe and interpret the results.   (100%)

FiO$_2$	0.21		
pH	7.49		
PCO$_2$	44	mmHg (5.9	kPa)
PO$_2$	98	mmHg (13	kPa)
HCO$_3^-$	34	mmol/L	
BE	+10		
O$_2$ Sats	99%		
Na$^+$	132	mmol/L	
K$^+$	1.8	mmol/L	
Cl$^-$	86	mmol/L	
Urea	3.7	mmol/L	
Creatinine	62	µmol/L	

24

ANSWER

1. Acid base status

Mild alkalaemia (pH 7.49)

Normocapnoea

Significantly elevated bicarbonate
> Metabolic alkalosis
>> expect CO_2 = 20 + 0.7 x 34
>> = 20 + 24
>> = 44 mmHg
>> therefore CO_2 is appropriate respiratory compensation
>> AG/DR N/A for alkalosis

Therefore diagnose primary metabolic alkalosis with appropriate respiratory compensation

2. Oxygenation

Normoxic on RA

A-a gradient   = 150 – 1.25 x 44 – 98
                       = 150 – 55 – 98
                       = 150 – 153
                       = -3

Not elevated. No V/Q mismatch.

3. Electrolytes

Mild hyponatraemia
Life threatening hypokalaemia
> for pH 7.49 expect K^+ = 5.0 - 1 x 0.5 = 4.5mmol/L
>> .: K^+ critically low when corrected for pH
>> High risk arrhythmias/respiratory failure
>> Needs immediate correction with high dose IV K^+

Moderate - Severe hypochloraemia
> Renal excretion

Renal indices normal
> No obvious renal failure

25

4. Interpret

ABG shows:
> Primary metabolic alkalosis with appropriate respiratory compensation
> Life threatening hypokalaemia
> Hypochloraemia

Metabolic alkalosis
> Given Hx presented must consider iatrogenic
>> Laxative abuse
>> Diuretic abuse
> → needs clarification with patient
> → check other electrolytes, UA and chemistry

> Other DDx include
>> Endocrinopathies (Cushing's, Bartter's Sx)
>> HCl loss – vomiting/anorexia nervosa

A-a gradient in kPa

A-a grad
$$= 21kPa - 1.25 \times 5.9 - 13$$
$$= 21 - 7.5 - 13$$
$$= 21 - 20.5$$
$$= 0.5kPa$$

Not elevated

26

COMMENTS

A 28 year old man presents with lethargy and dizziness. He has no medical history of note, and is a regular attendee of the local gym. He reports trying to lose weight over the last month, but denies any other symptoms.

Yet again the importance of the stem cannot be overstated in answering the ABG question. We are presented with a young patient with no medical history of note, which vastly decreases the pre-test probability of serious pathology. This makes the diagnosis of iatrogenic illness much more likely. This presentation has been included in this text as a situation many examinees may not have previously encountered.

The combination of metabolic alkalosis and hypokalaemia in a healthy patient with no pathology can represent either laxative abuse[11] or diuretic abuse[12].

The critically low K^+ is dealt with during the working of the question. It is central to the answer, probably pass/fail criteria, and therefore must be mentioned. Because it is mentioned in our standard working procedure it therefore will not be missed while an interpretation of the acid-base disorder is synthesized.

The A-a gradient comes out as negative in this answer. Although this should technically be a physiologic impossibility[13] there are several possible causes to consider including recent use of supplemental oxygen, mathematical or rounding error or measurement error in the sample. It is also worth noting that there are a significant number of inherent assumptions in the calculation of an A-a gradient, and that the application of a simple equation to a complex biological system is not perfect. Overall in the event of calculating a negative gradient under exam conditions candidates would be advised to check their mathematics, and if no error is found conclude that no V/Q mismatch exists.

[11] Oster JR, Masterson BJ and Rogers AI. Laxative abuse syndrome. *Am J Gastroenterol* 1980; 74(5): 451-58

[12] Papademetrious V. Diuretics, hypokalemia and cardiac arrhythmia: a 20-year controversy. *J Clin Hypertens* 2006; 8(2): 86 - 92

[13] Kanber GJ, King FW, Eschar YR and Sharp JT. The alveolar-arterial oxygen gradient in young and elderly men during air and oxygen breathing. *Am Rev Respir Dis* 1968; 97:376-81

PROBLEM 7

An 80 year old man attends your emergency department with a humeral head fracture. He is given intravenous morphine for pain. Thirty minutes later he is noted to be drowsy and unresponsive.

His vital signs are:

HR	60	/min
BP	150/75	mmHg
RR	6	/min
Sats	90%	RA
T	36.8	°C
GCS	7	

Arterial blood gas analysis is performed and the results are shown below.

Describe and interpret them.   (100%)

FiO$_2$	0.21		
pH	7.20		
PO$_2$	66	mmHg (8.8	kPa)
PCO$_2$	78	mmHg (10.4	kPa)
HCO$_3^-$	28	mmol/L	
BE	+2		
O$_2$ Sats	90%		
Na$^+$	136	mmol/L	
K$^+$	5.8	mmol/L	
Cl$^-$	100	mmol/L	
Glucose	6.5	mmol/L	

28

ANSWER

1. Acid base status

Moderate acidaemia (pH 7.20)

Significant hypercapnoea
  Respiratory acidosis
  Expect HCO_3^- for CO_2 78
    Acute = 24 + 4 = 28mmHg
    Chronic = 24 + 16 = 40mmHg

Mild-moderate increase in HCO_3^-
  Metabolic alkalosis
  Appropriate acute compensation for CO_2 78mmHg

Therefore diagnose acute respiratory acidosis with appropriate metabolic compensation.

2. Oxygenation

Critical hypoxia on room air
  Administer supplemental O_2
  Type II respiratory failure

A-a gradient = 150 − 1.25 x 78 − 66
      = 150 − 100 − 66
      = 150 − 166
      = - 16

Not elevated
No V/Q mismatch
  Hypoxia likely due to pure hypoventilation

3. Electrolytes

Normonatraemia
  Not cause of obtundation
Mild hyperkalaemia
  Expect K^+ for pH = 5 + 2 x 5 = 6.0 mmol/L
  K^+ level appropriate for pH
Normochloraemia

Mild hyperglycaemia
  Not cause for decreased GCS

29

4. Interpret

ABG shows acute respiratory acidosis with appropriate metabolic compensation.

Given clinical history strongly suggests opiate induced hypoventilation.
        Also consider
                Intracerebral insult
                        CVA
                        ICH/SAH
                        Undiagnosed head injury (ExDH, SDH)
                Secondary cause of low GCS
                        Sepsis

Vital signs show obtundation and hypoventilation
        Trial of reversal : naloxone
        If no success may need airway support
        No evidence of haemodynamic compromise as cause of low GCS

Likely iatrogenic insult
        Review case
        Quality assurance procedures
        Open disclosure to patient/family

A-a gradient in kPa

A-a grad
$$= 21 \text{ kPa} - 1.25 \times 10.4 - 8.8$$
$$= 21 - 13 - 8.8$$
$$= 21 - 21.8$$
$$= -0.8 \text{ kPa}$$

Not elevated

COMMENTS

An 80 year old man attends your emergency department with a humeral head fracture. He is given intravenous morphine for pain. Thirty minutes later he is noted to be drowsy and unresponsive.

Once again the stem sets up a fairly obvious answer. This patient is likely to have an iatrogenic suppression of respiration due to administration of opiates. Candidates should therefore expect an acutely compensated respiratory acidosis which is what the working indeed shows. The significance of a negative A-a gradient was discussed in problem 6.

Better candidates will also recognize the possibility of other causes of a decreased GCS with the given history of trauma. Again as an iatrogenic insult it is very important to mention relevant quality assurance issues. The principle of open disclosure is a key issue in the management of medical errors and should be mentioned for high marks[14]. The examiners here create potential opportunity to score higher marks by showing consultant level perspective.

[14] Iedema RA, Mallock NA, Sorensen RJ, Manias E, Tuckett AG, Williams AF et al. The national open disclosure pilot: evaluation of a policy implementation initiative. *MJA* 2008; 188(7):397-400

PROBLEM 8

An 80yo female with a history of rheumatoid arthritis presents with acute confusion. She lives alone and has not been seen for 2 days.

Her blood gas is shown below.

Describe and interpret the results.   (100%)

FiO$_2$	0.21		
pH	7.16		
PO$_2$	100	mmHg (13.3	kPa)
PCO$_2$	28	mmHg (3.7	kPa)
HCO$_3^-$	14	mmol/L	
BE	-8		
Na$^+$	120	mmol/L	
K$^+$	7.6	mmol/L	
Cl$^-$	100	mmol/L	
Glucose	4.0	mmol/L	
Lactate	1.2	mmol/L	

ANSWER

1. Acid-Base Balance

Moderate acidaemia
Significant hypocapnoea
      Respiratory alkalosis
Moderately low HCO_3^-
      Metabolic Acidosis
      Expect CO_2 = 8 + 1.5 x 14
                  = 29
      .: appropriate compensation
      AG   = 120 – 100 – 14
           = 6
      .: Dx NAGMA
      Delta R not calculable as AG <12 (ie N)

Therefore Dx NAGMA with appropriate respiratory compensation

2. Oxygenation

Gross normoxia

A-a gradient   = 150 – 1.25 x 28 – 100
                         = 150 – 35 – 100
                         = 115 – 100
                         = 15

Expect A-a for 80yo = 80/4 + 4 = 24
.: Dx N A-a gradient
      No V/Q mismatch

3. Electrolytes

Significant hyponatraemia
      Evaluate hydration status and medications (SSRI etc)
      Needs correction 0.5mmol/L.hr
Profound hyperkalaemia
      Expect K^+ pH 7.16   = 5.0 + 2.5 x 0.5
                              = 6.125mmol/L
      Therefore absolutely hyperkalaemic
      Check ECG
Normochloraemia
      NAGMA
Normoglycaemia
      Not cause of confusion

33

Normal lactate
    Less likely to be severely septic/shocked

4. Interpret

NAGMA plus hyponatraemia and hyperkalaemia
    Likely adrenal insufficiency
    Pt has RA – possible long term steroid use
    Supplemental steroids indicated

DDx for NAGMA in this setting: needs further evaluation for:

    Drugs – spironolactone, acetazolamide
    Fistula – uretoenteric, pancreaticoduodenal
    Renal failure – check renal indices and urine
    HCO_3^- loss – diarrhoea
    Panhypopituitarism
    Adrenal necrosis (Waterhouse-Friedrichson syndrome) if septic

Other things to consider
    Not seen 2 days
        Long lie syndrome
        Evaluate renal function and CK
    Drowsiness/confusion
        Evaluate other causes
        Septic screen, CT head

A-a gradient in kPa

A-a grad
    $= 21 \text{ kPa} - 1.25 \times 3.7 - 13.3$
    $= 21 - 4.5 - 13.3$
    $= 21 - 17.8$
    $= 3.2$

Not elevated (<4 kPa)

COMMENTS

An 80yo female with a history of rheumatoid arthritis presents with acute confusion. She lives alone and has not been seen for 2 days.

This stem is somewhat vague and represents a host of possibilities in this geriatric patient. It is an illustration of the need to both work through the question systematically, but also to consider the underlying possibilities and include these in the interpretation.

Systematic working of the question will reveal a NAGMA with associated hyponatraemia and hyperkalaemia, and the stem suggests the possibility of steroid use. Together this paints the picture of an acute adrenal crisis. Purists will argue that acute adrenal crisis complicated by hypokalaemia and hypernatraemia only occurs in primary adrenal insufficiency, because the pathophysiologic process is mediated by failure of mineral corticoid secretion[15]. Although classically steroid withdrawal only results in glucocorticoid deficiency, chronic adrenal fatigue and resultant acute adrenal crisis has been reported in patients undergoing long term glucocorticoid therapy[16]. The possibility of intercurrent sepsis and Waterhouse-Friedrichson syndrome (true adrenal crisis) should also be mentioned given the clinical picture, as should the possibility of excessive spironolactone use.

Strong candidates will also note the differential diagnosis for a NAGMA in their answer, and mention the possibility of a "long-lie" syndrome, which could also be associated with many of the electrolyte abnormalities described in the stem[17]. Thus renal function and a creatine kinase level should be urgently requested as part of the candidates "interpretation" of this question. Finally given the acute confusional state a CT brain should be considered.

[15] Burke CW. Adrenocortical insufficiency. *Clin Endocrinol Metab* 14(4): 947 -76
[16] Cronin CC, Callaghan N, Kearney PJ, Murnaghan DJ and Shanahan F. Addison disease in patients treated with glucocorticoid therapy. *Arch Intern Med* 1997; 157(4):456 -8.
[17] Szewczyk D, Ovadia P, Abdullah F and Rabinovici R. Pressure-induced rhabdomyolysis and acute renal failure. *J Trauma* 1998; 44(2): 384 - 8

PROBLEM 9

A 48 year old man presents complaining of nausea. He had a root canal two days ago and has consumed 26 compound paracetamol/codeine tablets in the last 24 hours for pain.

Selected investigations are shown below.

Describe and interpret the results.   (100%)

FiO$_2$	0.21		
pH	7.02		
PO$_2$	109	mmHg (14.5	kPa)
PCO$_2$	19	mmHg (2.5	kPa)
HCO$_3^-$	8	mmol/L	
BE	- 16		
O$_2$ Sats	100%		
Lactate	12.6	mmol/L	
Glucose	5.4	mmol/L	
Na$^+$	138	mmol/L	
K$^+$	6.3	mmol/L	
Cl$^-$	103	mmol/L	
AST	6500	IU/L	
ALT	5900	IU/L	
GGT	65	IU/L	
ALP	106	IU/L	
BR	50	mmol/L	

ANSWER

1. Acid base status

Life threatening acidaemia (pH 7.02)
Critical hypocapnoea
   Dx respiratory alkalosis

Severely low bicarbonate
   Critical metabolic acidosis
      expect CO_2 $= 8 + 1.5 \times 8$
           $= 19$ mmHg
     .: appropriate respiratory compensation

   AG $= 138 - 8 - 103$
          $= 27 \rightarrow$ RAGMA

   delta ratio $= (27-12)/(24-8)$
          $= 15/16$
          $= 1.0$.: Dx isolated RAGMA
          Fits with BE -16 for HCO_3^- 8

Therefore diagnose RAGMA and compensatory respiratory alkalosis.

2. Oxygenation

$pO_2 \rightarrow$ mild hyperoxia

A-a gradient $= 150 - 1.25 \times 19 - 109$
       $= 150 - 25 - 109$
       $= 150 - 136$
       $= 14$
Expect for 45yo male $45/4 + 4 = 11 + 4 = 15$
   .: Dx N A-a gradient
   No V/Q mismatch exists

3. Electrolytes

Normonatraemia
Moderate hyperkalaemia
   for pH 7.02 expect $5.0 + 4.0 \times 0.5 = 7.0$mmol/L
      K^+ mildly low for pH
      watch K^+ as pH corrected and correct if needed
Borderline hypochloraemia
   Electrical equilibration for RAGMA

37

Normoglycaemia
Grossly deranged liver function
    Hepatocellular picture
        Complicated by hyperbilirubinaemia
            Check conjugated fraction
        Likely paracetamol induced hepatotoxicity given history
Critical hyperlactataemia
    Likely cause of RAGMA:
        Type A - tissue hypoxia/shock
        Type B_1 sepsis/underlying disease (eg liver)
            Paracetamol Hepatotoxicity
        Type B_2 drug OD
            Possible drug toxicity: metformin, isoniazid, salycilates etc

4. Interpret

ABG shows:
    Critical Lactic acidosis plus acute hepatitis
    Appropriate respiratory compensation
        Mild relative hypokalaemia
            needs monitoring
        Probable paracetamol induced hepatotoxicity
            Obtain levels as proof
        Screen for other causes
            Hepatitis serology
            Ultrasound scan
        Needs **immediate** treatment with NAC
        ICU and gastroenterology consults
        Guarded prognosis → may require liver transplant

A-a gradient in kPa

A-a grad
    = 21kPa − 1.25 x 2.5 − 14.5
    = 21 − 3.3 − 14.5
    = 21 − 17.8
    = 3.2

Borderline elevated

COMMENTS

A 48 year old man presents complaining of nausea. He had a root canal two days ago and has consumed 26 compound paracetamol/codeine tablets in the last 24 hours for pain.

Even a brief read of the stem should enlighten the examinee as to the diagnosis of paracetamol hepatotoxicity. Over-ingestion of paracetamol leads to the formation of an electrophilic toxic metabolite[18], NAPQI, which causes centrilobular hepatic necrosis[19].

The question itself is a straightforward case of RAGMA associated with lactic acidosis, and should be worked as such. The deranged liver function indicates that this patient is at risk of life threatening paracetamol poisoning, and must be treated immediately with N-Acetyl Cysteine. Mentioning this is essential, even though the data regarding delayed administration indicates poorer prognosis[20].

For completeness sake strong candidates will comment on relevant minor issues, specifically correction of the K^+, the hyperbilirubinaemia and the need to measure a conjugated fraction, and the need to consider other causes of liver dysfunction. Finally, perspective is shown by obtaining relevant consults and mentioning the gravity of the prognosis.

---

[18] Manyike PT, Kharasch ED, Kalhorn TF and Slattery JT. Contribution of CYP2E1 and CYP3A to acetaminophen reactive metabolite formation. *Clin Pharmacol Ther* 2000; 67(3):275-82

[19] Lee WM. Drug-induced hepatotoxicity. *NEJM* 2003; 349(5):474-485

[20] Smilkstein MJ, Bronstein AC, Linden C, Augenstein WL, Kulig KW and Rumack BH. Acetaminophen overdose: a 48-hour intravenous N-acetylcysteine treatment protocol. *Ann Emerg Med* 1991; 20(10):1058-63

PROBLEM 10

A 45 year old woman presents to your department complaining of 6kg weight gain and lethargy over the last 3 weeks. She is not on medication and has been previously healthy, and says she thinks she has begun menopause.

Her vital signs are:

HR	80	/min
BP	165/90	mmHg
RR	12	/min
Sats	98%	RA
T	37.1	°C

An arterial blood gas is taken and the results are shown below.

Describe and interpret them.   (100%)

FiO$_2$	0.21		
pH	7.51		
PO$_2$	102	mmHg (13.6	kPa)
PCO$_2$	44	mmHg (5.9	kPa)
HCO$_3^-$	34	mmol/L	
BE	+10		
O$_2$ Sats	98%		
Na$^+$	152	mmol/L	
K$^+$	2.2	mmol/L	
Cl$^-$	122	mmol/L	
Glucose	11.5	mmol/L	

40

ANSWER

1. Acid base status

Moderate alkalaemia (pH 7.51)

Normocapnoea

Moderately elevated HCO_3^-
    Metabolic alkalosis
            expect CO_2 = 20 + 0.7 x 34
                                 = 20 + 23
                                 = 43 mmHg
        therefore CO_2 is appropriate compensation
    AG/DR not applicable for met alkalosis

Therefore diagnose primary metabolic alkalosis with appropriate respiratory compensation.

2. Oxygenation

Normoxic on RA
A-a gradient   = 150 − 1.25 x 44 − 102
                               = 150 − 55 − 102
                               = -7
Not elevated
Implies no V/Q mismatch

3. Electrolytes

Mild-moderate hypernatraemia
Critical hypokalaemia
    Expect K^+ for pH = 5 - 1 x 5 = 4.5 mmol/L
    Therefore grossly K^+ deficient
    High risk arrhythmia
    Needs urgent replacement and telemetry monitoring
Mild hyperchloraemia
Moderate hyperglycaemia
    Undiagnosed diabetes
    Steroid therapy
    Other endocrinopathy (eg Cushing's Sx)

4. Interpret

ABG shows:
>   Metabolic alkalosis with appropriate respiratory
>   compensation
>   Critical hypokalaemia
>   Moderate hyperglycaemia

Causes of metabolic alkalosis
>   Endocrinopathy
>>   Cushing's syndrome
>>>   Primary
>>>   Secondary
>>>   Paraneoplastic
>>>   Iatrogenic (check medication Hx for
>>>   steroids)
>>   Hyperaldosteronism
>>   Bartter's Sx
>   HCl loss
>>   Check if vomiting
>   Drugs
>>   Frusemide/other diuretics: less likely with weight
>>   gain

Overall suggests Cushing's Sx
>   Correlate clinically (striae, body habitus etc)
>   Endocrine consult
>>   Check other electrolytes and renal Fn
>>   Urine chemistry
>>   Consider further Ix (cortisol levels) and imaging
>>   (MR/CT brain and abdomen) in consult
>   Replace K$^+$
>>   High dose IV either KCl or KH_2PO_4
>>   Telemetry environment for same

A-a gradient in kPa

A-a grad
$$= 21kPa - 1.25 \times 5.9 - 13.6$$
$$= 21 - 7.5 - 13.6$$
$$= 21 - 21.1$$
$$= -0.1$$

Not elevated

COMMENTS

A 45 year old woman presents to your department complaining of 6kg weight gain and lethargy over the last 3 weeks. She is not on medication, has been previously healthy, and says she thinks she has begun menopause.

The stem suggests a non specific constellation of symptoms which are revealed to be associated with a metabolic alkalosis after working the question. The most likely cause is an endocrinopathy, and the triad of metabolic alkalosis, hypokalaemia and hypertension may represent either an excessive glucocorticoid (Cushing's) or mineralocorticoid (Conn's) syndrome[21,22].

In this case the clinical information given in the stem and the glucose intolerance favour a diagnosis of Cushing's syndrome, but both differentials should be mentioned, as well as the need to consult endocrinology. Giving a broader differential for the diagnosis of metabolic alkalosis is likely to earn more marks.

---

[21] Torpy DJ, Mullen N, Ilias I and Nieman LK. Association of hypertension and hypokalemia with Cushing's syndrome caused by ectopic ACTH secretion: a series of 58 cases. *Ann N Y Acad Sci* 2002; 970:134-44

[22] Stewart PM. Mineralocorticoid hypertension. *Lancet* 1999; 353(9161):1341-7

PROBLEM 11

A 58 year old woman on chemotherapy for breast cancer presents to your department feeling generally unwell.

Her vital signs are:

HR	105	/min
BP	110/65	mmHg
RR	28	/min
Sats	92%	6L
T	38.6	°C

Her ABG is shown below.

Describe and interpret the results.   (80%)

List further investigations you would consider.   (20%)

FiO_2	0.40		
pH	7.28		
PO_2	68	mmHg (9.0	kPa)
PCO_2	40	mmHg (5.3	kPa)
HCO_3^-	18	mmol/L	
BE	-6		
O_2 Sats	92%		
Na^+	141	mmol/L	
K^+	4.6	mmol/L	
Cl^-	106	mmol/L	
Glucose	5.8	mmol/L	

45

ANSWER

1. Acid base status

Mild acidaemia (pH 7.28)
Normocapnoea

Mildly low bicarbonate
       Mild metabolic acidosis

	expect CO_2	= 8 + 1.5 x 18
		= 35 mmHg

           .: inadequate respiratory compensation
           Relative respiratory acidosis

AG		= 141 – 18 – 106
		= 141 - 124
		= 17 → RAGMA

delta ratio		= (17-12)/(24-18)
		= 5/6
		= ~1.0 .: Dx isolated RAGMA
		Fits with BE –6 for HCO_3^- 18

Therefore diagnose RAGMA and compensatory respiratory alkalosis.

2. Oxygenation

pO_2 → significant hypoxia on supplemental O_2

A-a gradient		= 300 – 1.25 x 40 – 68
		= 300 – 50 – 68
		= 300 – 118
		= 172

Expect for 58yo female 58/4 + 4 = 14 + 4 = 18
        .: Dx elevated A-a gradient → Dx V/Q mismatch
               Concerning for LRTI/PE given sats and Hx
               ARDS (SIRS/aspiration)
               Interstitial lung disease
               Other respiratory membrane disease (eg APO)

3. Electrolytes

Normonatraemia
Normokalaemia
>for pH 7.28 expect 5.0 + 1 x 0.5 = 5.5mmol/L
>>.: K$^+$ very mildly low for pH
>>watch K$^+$ as pH corrected

Normochloraemia
Normoglycaemia
>DKA unlikely as cause of RAGMA

4. Interpret

ABG shows:
>RAGMA with inadequate respiratory compensation (acidosis) and V/Q mismatch

Febrile cancer pt on chemotherapy
>High risk sepsis
>Needs broad spectrum empiric Abx (gentamicin + pip/taz or cefepime)

RAGMA causes to consider:
>Lactate: check level
>>A – shock/tissue hypoxia
>>B$_1$ – sepsis, liver failure (mets)
>>>LRTI, urosepsis, line sepsis if vascular access device
>>B$_2$ – drugs: iron, isoniazid, paraldehyde, metformin, toxic alcohols
>Ketones
>>DKA unlikely with N BSL
>>Starvation
>>Dehydration
>Renal failure
>>Dehydration, chemotherapy

V/Q mismatch
>Febrile – concern RE LRTI. PE also more likely given Hx
>Ca

Therefore based on above, further Ix are:

Bedside:
      ECG
      Urine dip

Pathology:

FBC (WCC/Hb)
E/LFTs (renal fn, liver Fn, lipase, Ca^{++})
Lactate (sepsis)
Septic screen - Urine culture, BCs x 2 sets plus line, CXR
Drug levels if indicated
CSF m/c/s if concern RE meningitis

Imaging
CXR
CTPA if no cause for V/Q MM found
Other imaging as clinically indicated
      U/SS ?DVT
      CT abdomen or U/SS if abdominal focus suspected

A-a gradient in kPa

A-a grad
      = 40kPa – 1.25 x 5.3 – 9.0
      = 40 – 6.5 – 9.0
      = 40 – 15.5
      = 34.5

Grossly elevated

COMMENTS

A 58 year old woman <u>on chemotherapy</u> for breast cancer presents to your department feeling generally unwell.

Her vital signs are:

HR	*105*
BP	*110/65*
RR	*28*
Sats	*92% 6L*
T	*38.6°C*

Candidates will notice that the stem gives three very important pieces of information: fever and hypoxia in a patient on chemotherapy. This is a situation than mandates empiric antibiotics[23] and this should be made clear in the candidate's answer.

There are also two questions given. There is the usual "describe and interpret" prompt, but also a request for a "list" of further investigations. This should be regarded by the candidates as a request for a focused interpretation of the information given, rather than a need to generate an entirely new answer halfway through the question. In this case the clinical need will be to evaluate for a cause for the fever, and a cause for the elevated A-a gradients. Fellowship level candidates should generate an appropriate list of investigations without much difficulty.

[23] Flowers CR, Seidenfeld J, Bow EJ, Karten C, Gleason C, Hawley DK et al. Antimicrobial prophylaxis and outpatient management of fever and neutropenia in adults treated for malignancy: American Society of Clinical Oncology clinical practice guideline. *J Clin Oncol* 2013; 31(6):794-810

PROBLEM 12

A 26 year old man presents with dyspnoea and weakness. His parents mention that he has a history of "kidney problems" and that he has not been compliant with his medication.

His arterial blood gas on arrival is shown below.

Describe and interpret the results.   (100%)

FiO$_2$	0.21		
pH	7.08		
PO$_2$	110	mmHg (14.7	kPa)
PCO$_2$	19	mmHg (2.5	kPa)
HCO$_3^-$	7	mmol/L	
BE	-16		
O$_2$ Sats	100%		
Na$^+$	136	mmol/l	
K$^+$	1.8	mmol/L	
Cl$^-$	124	mmol/L	
Urea	4.7	mmol/L	
Creatinine	42	µmol/L	

ANSWER

1. Acid base status

Severe acidaemia (pH 7.08)

Profound hypocapnoea
     Respiratory alkalosis

Severely low bicarbonate
     Metabolic acidosis
               expect CO_2 = 8 + 1.5 x 7
                              = 18 mmHg
               therefore appropriate respiratory compensation

          AG   = 136 – 7 – 124
                              = 136 – 131
                              = 5 → NAGMA

          DR   n/a as AG < 12

Therefore diagnose NAGMA with appropriate respiratory
compensation

2. Oxygenation

pO_2 → mildly hyperoxic on RA
          probably secondary to hyperventilation due to respiratory
          compensation

A-a gradient   = 150 – 1.25 x 19 – 119
               = 150 – 25 – 119
               = 150 – 144
               = 6

Expect for 26yo male 26/4 + 4 = 4.5 + 4 = 8.5
          .: N A-a gradient. No V/Q mismatch

3. Electrolytes

Normonatraemia
Critical hypokalaemia
          for pH 7.08 expect K^+ = 5.0 + 3 x 0.5 = 6.5mmol/L
                    .: K^+ life threateningly low when corrected for pH
                    High risk arrhythmias/respiratory failure
                    Needs immediate correction with high dose IV K^+

51

Hyperchloraemia
        NAGMA
Renal indices normal
        No evidence renal failure as cause for NAGMA

Interpret

ABG shows:
        NAGMA with appropriate respiratory compensation
        Critical hypokalaemia

Findings consistent renal tubular acidosis
        Check urine pH (>5.3)
        Check other electrolytes, esp Ca^{++}
        pH may respond to HCO_3^- administration

Other causes of NAGMA to consider
        Hypoaldosteronism
                Adrenal insufficiency
                Spironolactone use
        Acetazolamide use
        NaCl administration
        Renal failure (unlikely with N indices)

Critical hypokalaemia
        Needs high dose K^+ replacement IV
        Telemetry/HDU involvement: high risk arrhythmia and
        respiratory failure

A-a gradient in kPa

A-a grad
        $= 21kPa - 1.25 \times 2.5 - 14.7$
        $= 21 - 3 - 14.7$
        $= 21 - 17.7$
        $= 3.3$ kPa

Borderline elevation.

COMMENTS

A 26 year old man presents with <u>dyspnoea and weakness</u>. His parents mention that he has a <u>history of "kidney problems"</u> and that he has <u>not been compliant with his medication</u>.

Candidates reading this stem may be perplexed as to the underlying implications. A good place to start with questions like this with no underlying clue to diagnosis in the stem is with systemic working to elucidate the major abnormalities. In this question these are NAGMA and hypokalaemia. In keeping with the practice of commenting on non acid-base issues during working of the question the importance of the hypokalaemia is identified and dealt with "in question". This is the likely cause of the patient's weakness[24], and will need to be urgently corrected. This is likely to be a pass/fail criteria.

Once the NAGMA is identified the list of differentials can be revisited to find a link with the information given in the stem. In this case renal loss of HCO_3^- is a likely to be relevant and should be articulated as such.

Both proximal tubular renal tubular acidosis (with or without Fanconi syndrome) and distal renal tubular acidosis are associated with NAGMA and hypokalaemia[25, 26], although the disturbances tend to be more severe with distal disease. Often patients with these conditions are maintained on bicarbonate therapy, presumably the medication with which this young man has not been compliant. It is very important to mention both the need to evaluate the other electrolytes, and to urgently replace the potassium in this patient as part of a sophisticated consultant level interpretation.

[24] Comi G, Testa D, Corneolio F, Comola M and Canal N. Potassium depletion myopathy: a clinical and morphological study of six cases. *Muscle Nerve* 1985; 8(1):17 - 21
[25] Sebastian A, McSherry E, and Morris RC Jr. Renal potassium wastin in renal tubular acidosis (RTA): its occurrence in types 1 and 2 RTA despite sustained correction of systemic acidosis. *J Clin Invest* 50(3): 667 - 78
[26] Batle D, Moorthi KM, Schlueter W and Kurtzman N. Distal renal tubular acidosis and the potassium enigma. *Semin Nephrol* 2006; 26(6): 471-8

PROBLEM 13

A 26 year old mechanic is brought in by ambulance after an overdose of an unknown substance. On arrival his vital signs are:

HR	106	/min
BP	80/50	mmHg
GCS	12	/min
sats	100%	RA
RR	34	/min

An arterial blood gas drawn on arrival in the resuscitation area is shown below.

Describe and interpret the results.   (100%)

FiO$_2$	0.21		
pH	7.15		
pO$_2$	116	mmHg (15 4	kPa)
pCO$_2$	20	mmHg (2.7	kPa)
HCO$_3^-$	8	mmol/L	
BE	-16		
Na	135	mmol/L	
K	6.5	mmol/L	
Cl	95	mmol/L	
glucose	6.2	mmol/L	
urea	8.1	mmol/L	
lactate	5.2	mmol/L	
measured osmolality	320	mosm/Kg	

54

ANSWER

1. Acid-Base Balance

Severe acidaemia (7.15)
Moderate hypocapnoea
               respiratory alkalosis
Severely low HCO_3^- and large -ve BE
               metabolic acidosis

expect CO_2	$= 8 + 1.5 \times 8$
	$= 20mmHg$

    .: diagnose appropriate respiratory compensation

AG	$= 135 - 95 - 8$
	$= 32$
Delta ratio	$= (32 - 12)/(24 - 8)$
	$= 20/16$
	$= 1.25$

    .: diagnose isolated RAGMA

Therefore Dx RAGMA with appropriate resp compensation

2. Oxygenation

Mild hyperoxaemia on room air

A-a grad	$= 150 - 1.25 \times 20 - 116$
	$= 150 - 25 - 116$
	$= 9$

Expect A-a 20yo $= 20/4 + 4 = 9$

N A-a gradient. No V/Q mismatch exists.

3. Electrolytes

Normonatraemia
Hyperkalaemia

expect K^+ for pH 7.15	$= 5 + 2.5 \times 0.5$
	$= 5 + 1.25$
	$= 6.25mmol/L$

    .: K^+ 6.5 is appropriate for pH
Mild hypochloraemia
    electrical equilibration RAGMA

Normoglycaemia
    not cause ALOC

Significant hyperlactataemia
> type A: hypoperfusion/shock
> type B_1: sepsis/liver failure. Check LFT
> type B_2: drugs
>> Metformin, iron, toxic alcohols, isoniazid, paraldehyde, paraquat

Hyperosmolar state
>> risk cerebral oedema
>> osmolar gap   = osm_c - osm_m
>>   = 320 - (2 x Na + urea + glucose)
>>   = 320 - 270 - 8 - 6
>>   = 50 - 14 = 36 ie high
> STRONGLY implies toxic alcohol ingestion
>> check urine for oxalate
>> check formic acid levels

4. Interpret

Overdose, young man, severe metabolic compromise
> compensated RAGMA plus raised osm gap
> suggests ethylene glycol or methanol intoxication
> urgent treatment indicated
>> antidotes: ethanol or fomepizole. Consult toxicology
>> haemodialysis likely indicated: consult ICU
> needs paracetamol level (co-ingestant)
>> psych input as required
> evaluate for other causes of lactate: septic screen, GI bleed etc

A-a gradient in kPa

A-a grad
> = 21kPa – 1.25 x 2.7 – 15.4
> = 21 – 3.4 – 15.4
> = 21 – 18.8
> = 2.2

Not elevated

COMMENTS

A 26 year old mechanic is brought in by ambulance after an overdose of an unknown substance.

Interestingly the stem here provides little clue as to the pathologic process examinees are expected to delineate. However a major clue may be garnered from examining the results candidates are expected to work:

measured osmolality   320   mosm/Kg

There are few reasons to provide a measured osmolality to candidates, unless the examiners expect calculation of an osmolar gap. In the setting of an overdose, this strongly implies the possibility of either methanol or ethylene glycol ingestion[27]. Candidates should expect a concurrent RAGMA with a high lactate. Systematic working of the question does indeed reveal these findings. Of note, calculating the DR reveals an answer >1.0. However this should not be interpreted as a concurrent metabolic alkalosis. Small deviations about 1 probably reflect an inherent inaccuracy in the method of calculation, and a metabolic alkalosis is unlikely unless the ratio exceeds 2.

To score highly candidates should demonstrate their understanding of the subject by reference to specific metabolites of methanol (formic acid)[28] and ethylene glycol (oxaloacetate)[29]. Relevant treatment by either inhibition of alcohol dehydrogenase inhibition[30] or haemodialysis should also be highlighted. The indications for haemodialysis are controversial, but include the presence of RAGMA, end organ damage or an osmolar gap >10[30]. Expert opinion should be sought.

Finally, good candidates, realizing that this is a potential episode of self harm should mention the need to test for co-ingestants and

[27] Lynd LD, Richardson KJ, Purssell RA, Abu-Laban RB, Brubacher JR, Lepik KJ et al. An evaluation of the osmole gap as a screening test for toxic alcohol poisoning. *BMC Emerg Med* 2008; 8:5
[28] D'Alessandro A, Osterloh JD, Chuwers P, Quinlan PJ, Kelly TJ and Becker CE. Formate in serum and urine after controlled methanol exposure at the threshold limit value. *Environ Health Perspect* 1994; 102(2):178-81
[29] Fraser AD. Clinical toxicologic implications of ethylene glycol and glycolic acid poisoning. *Ther Drug Monit* 2002; 24(2):232-8
[30] Barceloux DG, Krenzelok EP, Olson K and Watson W. American Academy of Clinical Toxicology Practice Guidelines on the Treatment of Ethylene Glycol Poisoning. Ad Hoc Committee. *J Toxicol Clin Toxicol* 1999; 37(5):537

seek appropriate input from toxicology, intensive care and psychiatry services.

PROBLEM 14

An 85 year old female presents with a left hip fracture. She has been on long term calcium and vitamin D supplements as osteoporosis prophylaxis. A venous blood gas is taken and the results are shown below.

Describe and interpret the results   (100%)

FiO_2	0.21		
pH	7.49		
PO_2	43	mmHg (5.7	kPa)
PCO_2	42	mmHg (5.6	kPa)
HCO_3^-	32	mmol/L	
BE	+8		
O_2 Sats	78%		
Na^+	138	mmol/L	
K^+	4.7	mmol/L	
Cl^-	102	mmol/L	
PO_4^{3-}	0.5	mmol/L	
Ca^{2+}	2.8	mmol/L	
Urea	12.8	mmol/L	
Creatinine	130	µmol/L	

59

ANSWER

1. Acid base status

Mild alkalaemia (pH 7.49)

Normocapnoea

Moderately elevated HCO_3^-
  Metabolic alkalosis
    expect CO_2 = 20 + 0.7 x 32
    = 20 + 21
    = 42 mmHg
    therefore CO_2 is appropriate compensation
  AG/DR not applicable for met alkalosis

Therefore diagnose primary metabolic alkalosis with appropriate respiratory compensation.

2. Oxygenation

Venous sample, therefore A-a gradient not applicable

3. Electrolytes

Normonatraemia
Normokalaemia
  Expect K^+ for pH = 5 - 1 x 5 = 4.5 mmol/L
  Therefore K^+ is roughly pH appropriate
Normochloraemia
Significant hypercalcaemia
Moderate hypophosphataemia
Elevated renal indices
  U:C ratio ~90
  Likely pre-renal
    Assess volume status
    Check medication history for contributors
    (ACEi/frusemide)

4. Interpret

ABG shows:
  Metabolic alkalosis with appropriate respiratory compensation
  Marked hypercalcaemia and hypophosphataemia
  Renal failure

60

Given clinical history this suggests Milk-Alkali Sx as a complication of long term Ca^{2+} therapy

    Treat IV fluids, monitor Ca^{2+}, consider diuretics/bisphosphonates

    Cease Ca^{2+} supplements

    Check ECG for QT prolongation

Other causes of high Ca^{2+}:

    Primary hyperparathyroidism

    Metastatic Ca/myeloma

        ?pathologic fracture: evaluate xrays

Other causes alkalosis:

    Endocrinopathies

        Cushing's Sx (less likely due to N electrolytes)

        Baarter's Sx

        Loop diuretics (unlikely with increased Ca/K)

        HCl loss

            Clarify if any vomiting

COMMENTS

An 85 year old female presents with a left hip fracture. She has been on long term calcium and vitamin D supplements as osteoporosis prophylaxis. A venous blood gas is taken and the results are shown below.

This stem again represents an unusual cue to an acid base disturbance, and candidates may wish to proceed first by systematically working the question. Doing so reveals a metabolic alkalosis complicated by hypercalcaemia and renal failure. This is the milk-alkali syndrome, caused by large amounts of calcium and ingestible alkali. Historically it was associated with milk and sodium bicarbonate treatment for peptic ulcer disease, but today is prevalent as a complication of the widespread use of calcium carbonate for osteoporosis prophylaxis[31]. Strong candidates will establish this diagnosis, as well as give a sound differential for causes of both hypercalcaemia and metabolic alkalosis as part of their interpretation.

---

[31] Picolos MK, Lavis VR and Orlander PR. Milk-alkali syndrome is a major cause of hypercalcaemia among non-end-stage renal disease (non-ESRD) inpatients. *Clin Endocrinol* 2005; 63(5): 566 - 76

PROBLEM 15

A 30 year old woman presents to ED after taking an overdose of an unknown substance three hours previously. The ambulance officers report that she is confused and reporting ringing in her ears.

Her vital signs are:

HR	112	/min
BP	99/62	mmHg
RR	34	/min
Sats	100%	RA
T	37.8	°C
GCS	14	

Arterial blood gas analysis is shown below.

Describe and interpret the results   (100%)

FiO^2	0.21		
pH	7.60		
PO_2	115	mmHg (15.3	kPa)
PCO_2	20	mmHg (2.7	kPa)
HCO_3^-	18	mmol/L	
BE	-4		
O_2 Sats	100%		
Na^+	135	mmol/L	
K^+	5.8	mmol/L	
Cl^-	98	mmol/L	
Glucose	5.8	mmol/L	
Lactate	0.6	mmol/L	
Urea	4.5	mmol/L	
Creatinine	0.06	mmol/L	

ANSWER

1. Acid base status

Profound alkalaemia (pH 7.60)

Profound hypocapnoea
> Respiratory alkalosis
> Expect HCO_3^- for CO_2 20
>> Acute $= 24 - 2 \times 2 = 20mmHg$
>> Chronic $= 24 - 2 \times 5 = 14mmHg$

Mild decrease in HCO_3^-
> Metabolic acidosis
>> Not consistent with either acute or chronic compensation for CO_2

AG	$= 135 - 18 - 98$
	$= 135 - 116$
	$= 19 \rightarrow$ RAGMA
DR	$= (19-12)/(24-18)$
	$= 7/6$
	$= {\sim}1$
	Implies isolated RAGMA

Therefore diagnose:
> acute respiratory alkalosis
> metabolic acidosis
>> partly compensation
>> partly RAGMA

2. Oxygenation

Hyperoxaemia on RA
> Secondary to gross hyperventilation
>> Implies central respiratory stimulation

A-a gradient
$= 150 - 1.25 \times 20 - 115$
$= 150 - 25 - 115$
$= 150 - 140$
$= 10$

Expect A-a gradient 30yo female $= 30/4 + 4 = 7.5 + 4 = 11.5$
Not elevated
No V/Q mismatch

3. Electrolytes

Normonatraemia
      Not cause of confusion
Mild hyperkalaemia
      Expect K^+ for pH = 5 - 2 x 0.5 = 4.0 mmol/L
      K^+ level elevated when corrected for pH
            Monitor telemetry as pH corrected
Mild hypochloraemia
      Renal equilibration for RAGMA
Normoglycaemia
      Not cause for decreased GCS/confusion
      DKA unlikely
Normal lactate
      Not cause for RAGMA
Normal renal indices
      Not cause for RAGMA

4. Interpret

ABG shows
      acute primary respiratory alkalosis with hyperoxaemia and
      some metabolic compensation
      concurrent RAGMA

Primary respiratory alkalosis in absence of hypoxia implies central
respiratory stimulation
      Most consistent with Aspirin overdose, given clinical Hx
            Also fits with RAGMA (N lactate and urea)
            Mx: alkalinize urine and IV fluid load
            Serial salicylate levels to Ax therapy
            Check paracetamol level (co-ingestant)
            Consult toxicology
            Psychiatry input RE suicide attempt

A-a gradient in kPa

A-a grad
      = 21 kPa – 1.25 x 20 – 15.3
      = 21 – 2.5 – 15.3
      = 21 – 17.8
      = 3.2

Borderline elevation

COMMENTS

A 30 year old woman presents to ED after taking an <u>overdose of an unknown substance</u> three hours previously. The ambulance officers report that she is <u>confused </u>and reporting <u>ringing in her ears</u>.

Again candidates are presented with an unknown overdose for working. However, again a clue is contained in the stem: the presence of tinnitus, classically a hallmark of aspirin overdose[32].

Working the question shows dual processes. There is a primary respiratory acidosis and a concurrent RAGMA with a normal lactate, which should immediately confirm the diagnosis for the candidate. This is a question which should engender a smile, because it a classic presentation of a core subject, and thus provides the opportunity for high marks.

Strong candidates will mention the lack of differentials for the RAGMA given a normal glucose and lactate. The fundamentals of aspirin overdose should be outlined. Alkalinization of the urine to enhance elimination is indicated[33]. Serum levels and the Done nomogram are not useful in isolation, unlike the Rumack-Matthews nomogram for paracetamol poisoning[34]. However, consecutively falling levels indicate an adequate response to treatment.

Finally, strong candidates will again mention the need to examine for co-ingestants and seek input from the mental health service.

---

[32] O'Malley, GF. Emergency department management of the salicylate-poisoned patient. *Emerg Med Clin North Am* 2007; 25(2):333-46

[33] Proudfoot AT, Krenzelok EP and Vale JA. Position paper on urine alkalinization. *J Toxicol Clin Toxicol* 2004; 42(1):1-26

[34] Dugandzic RM, Tierney MG, Dickinson GE, Dolan MC and McKinght DR. Evaluation of the validity of the Done nomogram in the management of acute salicylate intoxication. *Ann Emerg Med* 1989; 18(11):1186 - 90

THE HARDER PROBLEMS

"Don't panic"
Douglas Adams

PROBLEM 16

A 26 year old asthmatic with a history of ICU admissions presents to your department in acute respiratory distress. She has been treated by her GP for 2 days with bronchodilators and glucocorticoids. Her vital signs on arrival are:

HR	120	/min
BP	110/70	mmHg
RR	30	/min
Sats	92%	40% O_2
T	37.2	°C

Arterial blood gas analysis is undertaken in the resuscitation area.

Describe and interpret the results.   (100%)

FiO_2	0.4		
pH	7.32		
PO_2	65	mmHg (8.7	kPa)
PCO_2	48	mmHg (6.4	kPa)
HCO_3^-	18	mmol/L	
BE	-6		
O_2 Sats	92%		
Na^+	142	mmol/L	
K^+	3.2	mmol/L	
Cl^-	88	mmol/L	
Glucose	10.2	mmol/L	
Lactate	3.6	mmol/L	

68

ANSWER

1. Acid base status

Mild acidaemia (pH 7.32)

Mild hypercapnoea
  Respiratory acidosis
  Expect HCO_3^- for CO_2 48
    Acute = 24 + 1 = 25mmHg
    Chronic = 24 + 4 = 28mmHg

Mild decrease in HCO_3^-
  Metabolic acidosis
    expect CO_2 = 8 + 1.5 x 18
              = 8 + 27
              = 35 mmHg

AG   = 142 – 18 – 105
     = 142 – 123
     = 19 → RAGMA
DR   = (19 – 12)/(24 – 18)
     = 7/6
     = 1
     Implies isolated RAGMA

Therefore diagnose dual disturbance: primary respiratory acidosis and metabolic acidosis. No compensation.

2. Oxygenation

Critical hypoxia on supplemental oxygen.

A-a gradient
     = 300 – 1.25 x 48 – 65
     = 300 – 60 – 65
     = 300 – 125
     = 175

Grossly elevated ! (expect for 26yo female = 26/4 + 4 = ~10)

Gross V/Q mismatch
  Implies critical/life threatening asthma
  Also consider
    ARDS (SIRS/aspiration)

69

Interstitial lung disease
Other respiratory membrane disease (eg LRTI)
PE

3. Electrolytes

Normonatraemia
Mild hypokalaemia
    Possible intracellular shift if using salbutamol
    Expect K^+ for pH = 5 + 0.5 x 5 = 5.25 mmol/L
    Therefore significantly K^+ deficient on correction
    Replace K^+ and monitor ECG
Normochloraemia
Moderate hyperglycaemia
    Secondary to glucocorticoid administration
    Possible diabetes – check medical Hx

Moderate hyperlactataemia
    Likely due to salbutamol administration (B_2)
    Type A less likely with normal haemodynamics
    Type B_1 less likely with N temperature, but evaluate for
    sepsis

4. Interpret

This is a critically ill patient*

Asthmatic, and ABG shows the dual processes of
    Respiratory acidosis
    RAGMA

Respiratory acidosis
    VERY concerning given Hx of asthma
        Indicates critical/life threatening disease
        Hypercapnoeic asthma often precedes respiratory
        arrest
        Needs urgent treatment
            High dose bronchodilators, $MgSO_4$, CPAP,
            possible IPPV
        Involve intensive care service

RAGMA
    Hyperlactataemic
        Salbutamol
        Sepsis/shock less likely

Check to exclude other drug causes (Iron,
isoniazid, paraldehyde, toxic alcohols, metformin
etc)
Also consider other causes
Diabetic ketoacidosis
Hyperglycaemic: check med Hx and
urine/blood ketones
Uraemia
Check renal function

A-a gradient in kPa

A-a grad
$$= 40 \text{ kPa} - 1.25 \times 6.4 - 8.7$$
$$= 40 - 8 - 8.7$$
$$= 40 - 16.7$$
$$= 23.3 \text{ kPa}$$

Grossly elevated.

71

COMMENTS

A 26 year old asthmatic with a history of ICU admissions presents to your department in acute respiratory distress. She has been treated by her GP for 2 days with bronchodilators and glucocorticoids.

The stem again provides valuable information. This is a high risk patient[35], and candidates will need to add perspective to their answer.

Usually asthmatics present with hyperventilatory hypocapnoea. Here the presence of hypercarbia indicates severe asthma, and a PEFR <25% of the patient's norm[36]. Respiratory collapse is imminent. This is underlined by a comment during the evaluation of the A-a gradient to show the candidate's understanding of severity.

Several minor issues require comment in this blood gas, including the presence of a significant K^+ deficit, hyperglycaemia (presumably related to steroids) and hyperlactataemia (salbutamol therapy). Again these are included in the main body of working, so that the interpretation phase can be focused on the major issues.

Critically, in the interpretation phase the candidate must highlight the importance of the presentation. Phrases such as "this is a critically ill patient" are useful in this instance. An overview of relevant treatment is indicated. A full differential for the RAGMA can also be included should time allow.

---

[35] Dhuper S, Maggiore D, Chung V and Shim C. Profile of near-fatal asthma in an inner-city hospital. *Chest* 2003; 124(5): 1880-84

[36] Martin TG, Elenbaas RM and Pingleton SH. Use of peak expiratory flow rates to eliminate unnecessary arterial blood gases in acute asthma. *Ann Emerg Med* 1982 11(2):70

PROBLEM 17

A 3 year old boy presents with his parents after ingesting an unknown amount of iron tablets earlier in the day. He is vomiting profusely.

His venous blood gas is shown below.

Describe and interpret the results.　　　　　(100%)

FiO_2	0.21		
pH	7.16		
PO_2	65mmHg	(8.66	kPa)
PCO_2	25 mmHg	(3.3	kPa)
HCO_3^-	10mmol/L		
BE	- 8		
O_2 Sats	100%		
Lactate	5.6	mmol/L	
Glucose	5.4	mmol/L	
Na^+	146	mmol/L	
K^+	6.0	mmol/L	
Cl^-	90	mmol/L	
Urea	4	mmol/L	
Creatinine	70	µmol/L	

ANSWER

1. Acid base status

Significant acidaemia (pH 7.16)

Moderate hypocapnoea
    respiratory alkalosis

Significantly low bicarbonate
    Moderate - severe metabolic acidosis
       expect CO_2 = 8 + 1.5x10
                  = 23mmHg
        appropriate respiratory compensation

       AG   = 146 – 10 – 90
                       = 46 → RAGMA

       delta ratio   = (46 – 12)/(24-10)
                      = 34/14
                      >2 .: Dx co-existent metabolic acidosis and
                      metabolic alkalosis
                      Explains discrepancy of BE –8 for HCO_3^-
                      10

Therefore diagnose RAGMA, concurrent metabolic alkalosis and compensatory respiratory alkalsosis.

2. Oxygenation

pO_2 : venous sample
    cannot interpret A-a gradient

3. Electrolytes

Normonatraemia
Moderate hyperkalaemia
    note K^+ mildly high (6.0)
        for pH 7.16 expect 5.0 + 2.5 x 0.5 = 6.25mmol/L
        K^+ roughly appropriate for pH
        watch and replace k as pH corrected
Mild hypochloraemia
    electrical equilibrium/renal excretion RAGMA
    possible HCl loss from vomiting
Normal BSL
    DKA unlikely

Normal renal indices
> not cause for acidosis
> no renal failure complicating Fe^{3+} ingestion

Moderate hyperlactataemia
> Likely cause RAGMA:
>> Type A - tissue hypoxia : consider hypo-perfusion secondary shock/hypotension secondary vomiting
>>> Check volume status
>> Type B_1 sepsis/underlying disease (eg liver)
>> Type B_2 drug OD
>>> Consistent with significant Fe^{3+} OD
>>> Other drugs (metformin, isoniazid etc also possible)
>> Type B_3: inborn error of metabolism unlikely at this age

4. Interpret

3yo child: weight 14kg
> 60mg/kg = risk serious toxicity

VBG shows findings consistent with Fe overdose:
> Lactic acidosis
> Appropriate respiratory compensation
>> Needs quantification of dose ingested
>>> Historical: qualify type of Fe tabs
>>> Radiographic
>>> Iron levels
>> Desferrioxamine indicated (15mg/kg IVI titrated to *vin-rose urine*). Dialysis if develops renal failure.

Also metabolic alkalosis
> HCl loss from vomiting
> Possible co-ingestant (eg frusemide tabs)

Paediatric presentation
> Disclosure to parents
> Evaluate for NAI/neglect

COMMENTS

A 3 year old boy presents with his parents after ingesting an unknown amount of iron tablets earlier in the day. He is vomiting profusely.

Here candidates are given the diagnosis, and some background knowledge of iron toxicity will be required. A minimum diagnosis of a RAGMA should be sought.

Working this question reveals the issue is somewhat more complex. There is a triple acid-base disturbance. The key to identifying it is to take a systematic approach and work methodically through the CO_2 level, anion gap and delta ratio. Above all, when faced with a triple disturbance candidates should not panic. RAGMA and respiratory compensation fit with acute iron overdose. Profound hypochloraemia (mathematically driving up the AG in relation to the DR) reveals a concurrent metabolic alkalosis, most likely caused by HCl loss from vomiting. Returning to the information provided in the stem validates this theory. Other differentials (co-ingestant of a loop diuretic) can be inserted as they are thought of.

As this is a paediatric presentation the usual paediatric filters should apply. Although the formulation of iron preparations varies toxicity is likely at a dose >60mg/kg elemental iron ingested[37]. NAI and parental disclosure should be mentioned in all paediatric cases.

Indications for desferrioxamine include severe systemic compromise, presence of a RAGMA, iron levels above 90mcg/mL or abdominal xray showing a large number of ingested pills (iron tablets are radiopaque)[38]. In patients with renal failure dialysis can be used to remove the chelated molecules[39].

---

[37] Morris CC. Pediatric iron poisonings in the United States. *South Med J* 2000; 93(4):352 - 8
[38] Madiwale T, Liebelt E. Iron: not a benign therapeutic drug. *Curr Opin Pediatrics* 2006; 18(2): 174-9
[39] Carlsson M, Cortes D and Kanstrup T. Severe iron intoxication treated with exchange transfusion. *Arch Dis Child* 2008; 93(4):321-2

PROBLEM 18

A 68 year old diabetic woman presents with respiratory distress and severe thirst after a recent diarrhoeal illness. He daughter reports that her mother is on oral hypoglycaemics only. Her vital signs are:

HR	125	/min
BP	90/60	mmHg
T	37.8	°C
Sats	99%	RA
RR	38	/min

Her ABG from resus is shown below.

Describe and interpret the results. (100%)

FiO_2	0.21		
pH	6.98		
PO_2	119	mmHg (15.8	kPa)
PCO_2	14	mmHg (1.9	kPa)
HCO_3^-	3	mmol/L	
BE	- 21		
O_2 Sats	100%		
Lactate	8.6	mmol/L	
Glucose	11.4	mmol/L	
Na^+	138	mmol/L	
K^+	7.0	mmol/L	
Cl^-	114	mmol/L	
Urea	16.4	mmol/L	
Creatinine	150	µmol/L	

ANSWER

1. Acid base status

Life threatening acidaemia (pH 6.98)
Critical hypocapnoea
        Dx respiratory alkalosis

Critically low bicarbonate
        Critical metabolic acidosis
                expect CO_2 $= 8 + 1.5 \times 3$
                                        $= 11.5$ mmHg
             .: slight under respiratory compensation
             At this HCO_3^- physiologic limits are approached

        AG $= 138 - 3 - 114$
                              $= 21 \rightarrow$ RAGMA
        delta ratio $= (21\text{-}12)/(24\text{-}3)$
                              $= 9/21$
                              $= 0.5$.: Dx combined RAGMA and NAGMA

Therefore diagnose combined NAGMA/RAGMA and partially compensatory respiratory alkalosis.

2. Oxygenation

$pO_2 \rightarrow$ mild hyperoxia
        possibly secondary to elevated RR

A-a gradient $= 150 - 1.25{*}14 - 119$
                       $= 150 - 17 - 119$
                       $= 150 - 136$
                       $= 14$
Expect for 68yo female $68/4 + 4 = 17 + 4 = 21$
        .: Dx N A-a gradient
        No V/Q mismatch exists

3. Electrolytes

Normonatraemia
        Corrected for BSL $= 138 + (11.4 - 5)/3$
                                     $= 138 + 6/2 = 138 + 3$
                                     $= 141$mmol/L
Critical hyperkalaemia
        Check ECG
        for pH 6.98 expect $5.0 + 4.0 \times 0.5 = 7.0$mmol/L

78

K+ roughly appropriate for pH
watch K+ as pH corrected
Normochloraemia
Mildly elevated BSL
DKA possibly but less likely
(NB Not on insulin)
Abnormal renal indices
Both grossly elevated
U:C ratio ~100
Implies pre-renal failure
May contribute to NAGMA
Possible complication of
Diarrhoeal illness
Diabetes
Medication
Note Metformin is renally excreted
Moderate hyperlactataemia
Likely cause of RAGMA:
Type A - tissue hypoxia : consider hypo-perfusion
secondary shock/hypotension due vomiting
Type B$_1$ sepsis/underlying disease (eg liver)
Low grade Temperature
Type B$_2$ drug OD
Consistent with metformin toxicity
secondary to renal failure
Other drugs also possible

4. Interpret

ABG shows:
Critical Lactic acidosis (RAGMA)
Needs evaluation for cause
Septic screen
Check drug history RE metformin.
Renal failure (NAGMA)
Complicated by profound acidaemia
Haemodialysis may be indicated
In view of hypotension requires fluid load
Resuscitate to cap refill 2 secs/CVP
10mmHg
IDC and monitor UO
Evaluate for other causes of NAGMA
(endocrinopathy, HCO$_3$- loss due to diarrhoea)
Hyperkalaemia
Likely pH and renal failure related
Telemetry

79

Start insulin infusion RE increased BSL and K$^+$

Critically ill patient, guarded prognosis
Involve treating physicians
ICU consult

A-a gradient in kPa

A-a grad
= 21 kPa – 1.25 x 1.9 – 15.8
= 21 – 2.5 – 15.8
= 2.7

Not elevated for age

COMMENTS

A 68 year old diabetic woman presents with respiratory distress and severe thirst after a recent diarrhoeal illness. He daughter reports that her mother is on oral hypoglycaemics only.

This stem provides a large number of possibilities but no definitive guide as to the pathology underlying the arterial blood gas given.

Adequate candidates will be able to work through and define a mixed metabolic acidosis (RAGMA and NAGMA) with respiratory compensation. They can then proceed to give a list of differentials for the disorders and score a solid mark. Even though this question is of a slightly occult nature, and demonstrates multiple acid-base abnormalities, a systematic approach again will allow significant accumulation of marks.

Higher level candidates will recognize potential metformin toxicity in the context of acute renal failure. The cue here is the mention of oral hypoglyacaemics in the stem of the question. Metformin is renally excreted and thus will accumulate with a resultant lactic acidosis[40]. Because it has a small volume of distribution it will be removed by dialysis.

---

[40] Graham GG, Punt J, Arora M, Day RO, Doogue MP, Duong JK et al. Clinical pharmacokinetics of metformin. *Clin Pharmacokinet* 2011 50(2):81-98

PROBLEM 19.

A 65 year old long term smoker presents via the ambulance with fevers and dyspnoea. His observations are:

HR	105	/min
BP	90/60	mmHg
T	38.5	°C
RR	32	/min
Sats	89%	28% O_2

A blood gas is performed.

Describe and interpret the results.　　　　　　　(100%)

FiO_2	0.28		
pH	7.20		
PO_2	66	mmHg (8.8	kPa)
PCO_2	68	mmHg (9	kPa)
HCO_3^-	20	mmol/L	
BE	+4		
Na^+	145	mmol/L	
K^+	6.0	mmol/L	
Cl^-	100	mmol/L	
Urea	12	mmol/L	
Creatinine	100	µmol/L	
Lactate	2.6	mmol/L	

82

ANSWER

1. Acid-Base Balance

Moderate acidaemia
       Significant hypercapnoea
       Respiratory acidosis
            expect HCO_3^-:
                  acute to be 24 + 3 = 27mmol/L
                  chronically to be 24 + 4 x 3 = 36mmol/L
Mild decrease HCO_3^-
       Metabolic Acidosis
            Expect CO_2 = 8 + 1.5 x 20
                        = 38mmHg
            Anion Gap   = 145 − 20 − 100
                        = 25
            Delta Ratio   = (25 − 16)/(24 − 20)
                        = 9/5 >2.0
                Implies pre-extant or co-extant metabolic
                alkalosis

Therefore diagnose triple disturbance of respiratory acidosis, metabolic acidosis and metabolic alkalosis (likely compensation for chronic CO_2 retention)

2. Oxygenation

Significantly hypoxic
A-a gradient 29% O_2
              = 225 − 1.25x68 − 66
              = 225 − 85 − 66
              = 225 − 151
              = 74

Expect 65yo man A-a = 65/4 + 4 = 20
Raised A-a grad implies V/Q Mismatch
       LRTI/APO
       ARDS
       Chronic resp membrane disease
       PE

3. Electrolytes

Normonatraemia
Mild hyperkalaemia
       For pH 7.20 expect K^+ = 5.0 + 2 x 0.5 = 6.0mmol/L

83

K$^+$ appropriate for pH
Normochloraemia
Mild elevation of renal indices
    U:C ratio > 100
    Implies pre-renal failure
Mild hyperlactataemia (<4.0mmol/L)
    Type A – dehydration/hypo-perfusion (renal failure)
    Type B$_1$ – sepsis (febrile)
    Type B$_2$ – check medications (metformin, iron etc)

4. Interpet

Critically ill patient
Respiratory acidosis
    Likely acute on chronic given co-extant metabolic alkalosis
    Smoker – likely COPD
    BiPAP indicated until resolution of acute fraction of
    respiratory acidosis
RAGMA plus high lactate
    Resuscitate, fluids, bronchodilators, steroids if wheezy
    Broad spectrum ABx
    Needs full septic screen
Metabolic alkalosis
    Likely chronic if Hx COAD
    Consider other causes eg loop diuretics

A-a gradient in kPa

A-a grad
    $= 28 - 1.25 \times 9 - 8.8$
    $= 28 - 11.5 - 8.8$
    $= 28 - 20.3$
    $= 7.7$ kPa

Elevated

COMMENTS

A 65 year old long term smoker presents via the ambulance with fevers and dyspnoea.

Although this question may seem very convoluted, with a triple acid-base disturbance, it illustrates the importance of noting the clues given in the stem. In a long term smoker with fever, dyspnoea and hypotension the obvious picture is one of an infective exacerbation of COPD.

The key to understanding the triple disturbance is the delta-ratio of >2.0. In other words the fall in anion gap has been more than twice the fall in bicarbonate expected. The implication is of a pre-extant metabolic alkalosis, most likely in a smoker as part of the picture of chronic type II respiratory failure. Good candidates might mention another possibility for the metabolic alkalosis (in this case the use of diuretics) to show completeness of knowledge.

Neither pH nor HCO_3^- correct for each other, and this makes it very difficult to delineate the exact degree of acute respiratory acidosis. The co-existing metabolic alkalosis however makes it quite likely that there is an element of chronic CO_2 retention.

Finally, to draw a line under the clinical information given, a good candidate will mention administration of medical therapy and a trial of bi-level non invasive ventilation, which has a treatment benefit for these patients[41].

---

[41] Ram FS, Picot J, Lightowler J and Wedzicha JA, Non-invasive positive pressure ventilation for treatment of respiratory failure due to exacerbations of chronic obstructive pulmonary disease. *Cochrane Database Syst Rev.* 2004.

PROBLEM 20

Your intern asks you to a review a patient in the short stay ward. She is a 26 year old previously well woman, admitted with dehydration because of gastroenteritis. Overnight she has had 6L of 0.9% NaCl, and she is now complaining of dyspnoea and orthopnoea.

Her ABG results are shown below.

Describe and interpret the results.   (70%)

Outline your further actions.   (30%)

FiO_2	0.21		
pH	7.15		
PO_2	62	mmHg (8.26	kPa)
PCO_2	50	mmHg (6.7	kPa)
HCO_3^-	16	mmol/L	
BE	-8		
O_2 Sats	91%		
Na^+	151	mmol/L	
K^+	5.2	mmol/L	
Cl^-	126	mmol/L	
Urea	5.7	mmol/L	
Creatinine	60	µmol/L	

ANSWER

1. Acid base status

Moderate acidaemia (pH 7.15)

Hypercapnoea
        Respiratory acidosis
                Expect HCO_3^- for CO_2 50:
                        Acute   = 24 + 1 x 1 = 25
                        Chronic   = 24 + 1 x 4 = 29
Moderately low bicarbonate
        Metabolic acidosis
                expect CO_2 = 8 + 1.5 x 16
                                        = 32 mmHg
        AG   = 151 – 16 – 126
                = 151 – 142
                = 9 → NAGMA
        DR   = (9-12)/(24-18)
                = <0
        .: Dx isolated NAGMA
        Fits with BE –8 for HCO_3^- 16

Therefore diagnose dual process: combined respiratory acidosis and NAGMA.

2. Oxygenation

pO_2 → significant hypoxia on RA
                        needs supplemental O2

A-a gradient   = 150 – 1.25 x 50 – 62
                = 150 – 62 – 62
                = 150 – 124
                = 26

Expect for 26yo female 26/4 + 4 = 4.5 + 4 = 8.5
                .: Dx elevated A-a gradient → Dx V/Q mismatch
                        Concerning for APO given Hx of large volume
crystalloid
                        ARDS (SIRS/aspiration)
                        Interstitial lung disease
                        Other respiratory membrane disease (eg LRTI)
                        PE

3. Electrolytes

Mild hypernatraemia
> Secondary Na^+ load

Mild hyperkalaemia
> for pH 7.15 expect $5.0 + 1.5 \times 0.5 = 5.75$mmol/L
> > .: K^+ very mildly low for pH
> > watch K^+ as pH corrected

Hyperchloraemia
> NAGMA
> Cl^- load from NaCl

Renal indices normal
> No evidence pre-renal failure

4. Interpret

ABG shows:
> Combined respiratory acidosis, V/Q mismatch and NAGMA

Respiratory acidosis + V/Q mismatch
> Iatrogenic APO (8L crystalloid!)
> Other pulmonary disease
> > Pneumonia
> > PE
> > ARDS
> Needs CXR and evaluation for same

NAGMA
> Likely iatrogenic – excessive NaCl administration
> Other causes
> > Drugs – acetazolamide, spironolactone
> > Addison's – less likely given elevated Na^+
> > Renal failure unlikely with N indices

Therefore based on above, further actions are:

Ix and Rx patient
> CXR, volume status Ax, septic screen if warranted
> Treat APO – CPAP, diuresis, supplemental O_2
> Arrange admission

Iatrogenic error
> Investigate systems/process/individual
> Educate staff – individuals, review at M+M
> Apologize to patient: full and open disclosure

88

Documentation of same
Manage risk : brief director, medicolegal organization

Ongoing quality assurance and feedback for any changes made

A-a gradient in kPa

A-a grad
$$= 21kPa - 1.25 \times 6.7 - 8.3$$
$$= 21 - 8.5 - 8.3$$
$$= 21 - 16.8$$
$$= 4.2$$
Elevated

COMMENTS

Your intern asks you to a review a patient in the <u>short stay ward</u>. She is a <u>26 year old previously well woman</u>, admitted with <u>dehydration</u> because of gastroenteritis. Overnight she has had <u>6 litres of 0.9% NaCl</u>, and she is now complaining of <u>dyspnoea and orthopnoea</u>.

Her ABG results are shown below.

Describe and interpret the results	*(70%)*
Outline your further actions	*(30%)*

This question exemplifies a hard VAQ, because it occurs in two parts. Time pressure may be an issue. Candidates are asked to describe and interpret, as well as outline their further actions. However, the presence of a second part to the question asking for further actions should reassure candidates. It implies that the ABG will be diagnostic, as there are specific details the examiners will subsequently look for. Often the "part B" of a question is a natural extension of the interpretation phase of the question.

Here, the stem clearly suggests iatrogenic pulmonary oedema. Normal saline has a pH of around 5.5[42] and will result in a NAGMA if administered in excessive amount. Thus candidates could expect acute pulmonary oedema manifesting as respiratory acidosis with an elevated A-a gradient, and possibly a NAGMA. It is important, as always, to mention a list of relevant differentials.

When answering a question asking for an outline of further actions, broad brush subject headings are very helpful. In this case a consultant would ensure the welfare of the patient, manage an iatrogenic error, apologize to the patient and evaluate the ED system to ensure the error does not recur.

---

[42] Reddi B. Why is saline so acidic (and does it really matter?). *Int J Med Sci* 2013; 10(6): 747 - 750

PROBLEM 21

A 16 year old girl presents with her parents with thirst and dyspnoea. Her parents report periods of excessive vomiting over the last six days, accompanied by significant weight loss. Her BMI is 16.

Her venous blood gas is shown below.

Describe and interpret the results.   (100%)

FiO$_2$	0.21		
pH	7.42		
PO$_2$	98	mmHg (13	kPa)
PCO$_2$	35	mmHg (4.7	kPa)
HCO$_3^-$	18	mmol/L	
BE	+6		
O$_2$ Sats	99%		
Na$^+$	138	mmol/L	
K$^+$	2.8	mmol/L	
Cl$^-$	66	mmol/L	
Mg$^+$	0.7	mmol/L	
PO$_4^-$	0.5	mmol/L	
Albumin	28	g/L	
Glucose	5.4	mmol/L	
Urea	1.2	mmol/L	
Creatinine	32	μmol/L	
Lactate	0.6	mmol/L	

ANSWER

1. Acid base status

pH normal – not acidaemic or alkalaemic

Normocapnoea

Mildly decreased HCO_3^-
    Metabolic acidosis
        expect CO_2 = 8 + 1.5 x 18
                         = 8 + 27
                         = 35 mmHg
        therefore CO_2 is appropriate respiratory compensation
    AG   = 138 – 18 – 66
              = 138 – 84
              = 54 → RAGMA
    DR   = (54 – 12)/(24-18)
              = 32/6
              = ~5
        Identifies concurrent metabolic alkalosis occurring
        Explains discrepancy between BE +6 and HCO_3^- 18

Therefore diagnose mixed metabolic picture: RAGMA and metabolic alkalosis occurring concurrently

2. Oxygenation

VBG – cannot interpret oxygenation status

3. Electrolytes

Normonatraemia

Mild- moderate hypokalaemia
    No correction for pH

Critical hypochloraemia
    Renal excretion or GIT loss (vomiting)

Mild hypomagnesaemia
    Nutritional/renal loss

Moderate hypophosphataemia
    May be nutritional or due to hyperventilation

Moderate hypoalbuminaemia
> Broad DDx: nutrition, inflammation, oedema, ca

Markedly decreased renal indices
> Poor muscle mass (BMI 16)
>> Decreased production of metabolites

Normoglycaemia
> DKA unlikely as cause of RAGMA

Normal lactate
> Unlikely to be cause of RAGMA

4. Interpret

ABG shows:
> Mixed RAGMA and alkalosis
> Multiple electrolyte abnormalities: low K^+, Mg^+, PO_4^{3-}
> Evidence of malnutrition/low muscle bulk

Clinical Hx implies significant eating disorder
> Annorexia Nervosa/Bulaemia Nervosa

Complicated by:
> RAGMA
>> Possible starvation ketoacidosis
>>> Check urinary/blood ketones
>> Other causes less likely with N glucose, renal indices and lactate. If evidence sepsis perform septic screen
> Metabolic alkalosis
>> Loss of HCl secondary vomiting
>>> Explains critical hypochloraemia
>> Also consider diuretic abuse, endocrinopathies
> Electrolyte abnormalities
>> High risk refeeding syndrome
>> Need prophylactic replacement and cardiac monitoring
>> Can confirm malnutrition with pre-albumin assay

Requires admission
> Full disclosure to parents
> Involve ICU, psychiatry, eating disorders unit

COMMENTS

A 16 year old girl presents with her parents with thirst and dyspnoea. Her parents report periods of excessive vomiting over the last six days, accompanied by significant weight loss. Her BMI is 16.

The stem provided to candidates gives a strong indicator that the patient has an eating disorder. Starvation ketoacidosis is a rare but important cause of RAGMA, and is often accompanied by multiple electrolyte abnormalities[43]. The role of excessive HCl loss with a resultant metabolic alkalosis has already been discussed. Here it explains the widely distorted delta ratio.

Hypoalbuminaemia and low plasma creatinine may be reflective of decreased muscle bulk (as well as many other pathologies). The hallmark of refeeding syndrome is hypophosphataemia and this may cause acute respiratory failure. Thus early replacement of all major electrolytes is warranted.

Strong candidates will again note the role of other services in this patient's care, particularly the need for the involvement of an eating disorder psychiatrist. Other differentials for the major metabolic findings should also be mentioned.

---

[43] NiBhraonain S and Lawton L. Chronic malnutrition may in fact be an acute emergency. *J Emerg Med* 2013; 44: 72-4

PROBLEM 22

A 65 year old female with a history of bowel resection for colorectal cancer presents to the emergency room with a distended, painful abdomen and profuse vomiting.

Her vital signs are:

HR	120	/min
BP	99/55	mmHg
RR	22	/min
Sats	92%	40% O_2
T	38.7	°C

Her arterial blood gas is shown below.

Describe and interpret the results. (100%)

FiO_2	0.40		
pH	7.46		
PO_2	66	mmHg (8.8	kPa)
PCO_2	68	mmHg (9	kPa)
HCO_3^-	37	mmol/L	
BE	+12		
O_2 Sats	92%		
Na^+	138	mmol/L	
K^+	4.8	mmol/L	
Cl^-	72	mmol/L	
Lactate	1.2	mmol/L	
Glucose	2.6	mmol/L	

95

ANSWER

1. Acid base status

Mild alkalaemia (pH 7.46)

Hypercapnoea
>Respiratory acidosis

Significantly elevated HCO_3^-
>Metabolic alkalosis
>>expect CO_2 $\quad$ = 20 + 0.7 x 37
>>$\quad$ = 20 + 27
>>$\quad$ = 47 mmHg
>>therefore CO_2 is more than expected as appropriate compensation
>AG/DR not applicable for met alkalosis

Therefore diagnose primary metabolic alkalosis, and concurrent (non compensatory) respiratory acidosis

2. Oxygenation

Critically hypoxic on supplemental oxygen

A-a gradient $\quad$ = 300 − 1.25 x 68 − 66
$\quad$ = 300 − 85 − 66
$\quad$ = 300 − 151
$\quad$ = 149

Expect A-a gradient 65yo female = 65/4 + 4 = 16 + 4 = 20

Dx raised A-a gradient and V/Q mismatch
>ARDS (SIRS/aspiration) due to vomiting
>Interstitial lung disease
>Other respiratory membrane disease (eg LRTI/APO)
>PE

3. Electrolytes

Normonatraemia
Normokalaemia
>Expect K^+ for pH = 5 - 0.5 x 5 = 4.75 mmol/L
>Therefore K^+ is pH appropriate
Profound hypochloraemia
>GIT loss due to vomiting

96

Significant hypoglycaemia
>Needs correction with 50ml 50% dextrose IV

Normal lactate
>Severe sepsis/mesenteric ischaemia less likely

4. Interpret

ABG shows:
>Metabolic alkalosis and respiratory acidosis as two distinct entities

Metabolic alkalosis
>Hypochloraemic
>Hx vomiting and surgery – implies bowel obstruction with HCl loss
>Also evaluate other causes
>>Bartter's/Cushing's syndrome
>>Drugs: loop diuretics
>>Milk-alkali syndrome

Respiratory acidosis and hypoxia
>Type II respiratory failure
>V/Q mismatch – given fever, and Hx vomiting concern RE aspiration
>>CXR
>>Early treatment for aspiration LRTI
>>Search for other causes: BC x 2 and Urine m/c/s
>>Other causes of V/Q mismatch as mentioned earlier
>>CTPA if clinically indicated to exclude PE

Needs treatment for bowel obstruction + LRTI
>IV fluid titrated to HR < 100 and SBP > 100mmHg
>Abx for source of sepsis
>IDC, surgical input plus AXR or CT abdomen
>Analgesia/antiemetics
>Replace glucose

A-a gradient in kPa

A-a grad
$$= 40 \text{ kPa} - 1.25 \times 9 - 8.8$$
$$= 40 - 11.5 - 8.8$$
$$= 40 - 20.3$$
$$= 19.7$$

Grossly elevated

COMMENTS

A 65 year old female with a history of bowel resection for colorectal cancer presents to the emergency room with a distended, painful abdomen and profuse vomiting.

This question represents a double acid base disturbance of metabolic alkalosis and respiratory acidosis. The respiratory acidosis is more severe than would be expected as calculated compensation, and thus (in conjunction with a raised A-a gradient) is likely to represent independent pathology. Strong candidates will describe both abnormalities with appropriate differential diagnoses, and then link them back to the clinical cues provided in the stem. It is highly likely this lady has a bowel obstruction with vomiting complicated by HCl loss and aspiration.

Again, working through the question in a systematic fashion is key to achieving a high mark.

PROBLEM 23

A 60 year old patient presents to ED with pleuritic chest pain and haemoptysis. He has a history of DVT three years ago and is currently on aspirin.

His vital signs are:

HR	135	/min
BP	82/45	mmHg
RR	39	/min
Sats	88%	15L NRB
T	36.8	°C

Urgent arterial blood gas analysis is shown below.

Describe and interpret the results   (100%)

FiO$_2$	0.6		
pH	7.10		
PO$_2$	55	mmHg (7.3	kPa)
PCO$_2$	55	mmHg (7.3	kPa)
HCO$_3^-$	16	mmol/L	
BE	-8		
O$_2$ Sats	88%		
Na$^+$	139	mmol/L	
K$^+$	6.0	mmol/L	
Cl$^-$	103	mmol/L	
Lactate	8.7	mmol/L	

ANSWER

1. Acid base status

Profound acidaemia (pH 7.10)

Moderate hypercapnoea
        Respiratory acidosis
        Expect HCO_3^- for CO_2 60
                Acute   = 24 + 2 = 26mmHg
                Chronic = 24 + 8 = 32mmHg

Moderate decrease in HCO_3^-
        Metabolic acidosis
                expect CO_2 = 8 + 1.5 x 16
                                 = 8 + 22
                                 = 30 mmHg
      AG   = 139 – 16 – 103
                = 139 – 119
                = 20 → RAGMA
      DR   = (20 – 12)/(24 – 16)
                = 8/8
                = 1
                Implies isolated RAGMA

Therefore diagnose dual disturbance: primary respiratory acidosis and metabolic acidosis. No compensation.

2. Oxygenation

Life threatening hypoxia on supplemental oxygen.

A-a gradient   = 450 – 1.25 x 60 – 55
                                         = 450 – 75 – 55
                                         = 450 – 130
                                         = 320
Grossly elevated!
expect for 60yo male = 60/4 + 4 = 19

Gross V/Q mismatch
        Given Hx and haemodynamic instability implies massive PE
        Also consider
                ARDS (SIRS/aspiration)
                Interstitial lung disease

Other respiratory membrane disease (eg LRTI,
APO)

3. Electrolytes

Normonatraemia
Mild hyperkalaemia
Expect K^+ for pH = 5 + 3 x 5 = 6.5 mmol/L
Actually mildly hypokalaemic for pH
Monitor K^+ as pH corrected
Normochloraemia

Significant hyperlactataemia
Probably type A: gross tissue hypo perfusion (obstructive
shock)
Other causes
Type B_1 – sepsis, liver failure (less likely given Hx)
Type B_2 – drugs (salbutamol, isoniazid, iron,
paraldehyde, toxic alcohols etc)

4. Interpret

This is a critically ill patient*

ABG shows the dual processes of
Respiratory acidosis with profound V/Q mismatch
RAGMA + hyperlactaemia

Implies massive PE, haemodynamically unstable patient, peri-arrest
May be too unstable for CTPA or ECHO
Indication for urgent thrombolysis (or embolectomy)
Informed consent if possible
Notify respiratory medicine/Cardiothoracic Surgery
urgently
Clinically/radiographically (mobile CXR/FAST-E) exclude
tension Ptx and tamponade as causes of obstructive shock
prior to lysis
ECG – evaluate for R heart strain/arrythmia
ECHO could be used to confirm R heart strain prior to lysis

Causes of V/Q mismatch as noted earlier. Evaluate for these.

Other causes of RAGMA to consider (much less likely in this patient)

Diabetic ketoacidosis
Check BSL/med Hx
Uraemia
Check renal function

A-a gradient in kPa

A-a grad
= 60 kPa – 1.25 x 7.3 – 7.3
= 60 – 9 – 7.3
= 60 – 16.3
= 43.7

Grossly elevated

103

COMMENTS

A 60 year old patient presents to ED with pleuritic chest pain and haemoptysis. He has a history of DVT three years ago and is currently on aspirin.

The information provides the likely diagnosis. It is highly probable that this patient has suffered an episode of pulmonary venous thromboembolism. Candidates should note the observations given in this case:

HR	*135*
BP	*82/45*
RR	*39*
Sats	*88% 15L*
T	*36.8°C*

This patient is clearly peri-arrest, and could reasonably be expected to have systemic underperfusion manifesting as a lactic acidosis. Systemic working will reveal the two independent processes occurring, namely respiratory acidosis with a massive A-a gradient and a RAGMA, but the key here is to show perspective. The patient is critically ill, and there are multiple indications given for consideration of thrombolytic therapy[44]. The role of echocardiography in urgent assessment has long been established[45] and strong candidates will mention this as an important risk stratification tool. The importance of excluding other causes of obstructive shock prior to any intervention should also be made clear.

---

[44] Goldhaber SZ. Modern treatment of pulmonary embolism. *Eur Respir J* 2003; Suppl 35:22s
[45] Come PC. Echocardiographic evaluation of pulmonary embolism and its response to therapeutic interventions. *Chest* 1992; 101:151s

PROBLEM 24

A 40 year old man is brought by ambulance to your emergency department after a fire at a factory. He has no airway compromise or evidence of cutaneous burns, but he is GCS 12 and acutely confused. His blood pressure is 120/80 and his pulse oxygen saturation is 99%.

A blood gas is shown below.

Describe and interpret the results.　　　　　　(100%).

FiO_2	0.21		
pH	6.96		
PO_2	82mmHg	(10.9	kPa)
PCO_2	58mmHg	(7.7	kPa)
HCO_3^-	5mmol/L		
BE	-19		
HbCO	26%		
O_2 Sats	91%		
Lactate	16mmol/L		
Glucose	5.4mmol/L		
Na^+	136mmol/L		
K^+	4.0mmol/L		
Cl^-	96mmol/L		
Urea	5mmol/L		
Creatinine	0.09mmol/L		

105

ANSWER

1. Acid base status

Critical acidaemia (pH 6.96)

Moderate hypercapnoea
    Respiratory acidosis
    hypo ventilation 2ndary head injury/loc/seizure
    HCO_3^- compensation acute should = 26mmol/L

Critically low bicarbonate and BE -19
        critical metabolic acidosis
        expect CO2   = 8 + 1.5x5
                      = 15mmHg
    AG = 35
    delta ratio   = (35-12)/(24-4)
                   = 23/19 = 1.1 (<2.0) · isolated RAGMA

Therefore diagnose respiratory acidosis and concurrent RAGMA.
Dual process, no compensation.

2. Oxygenation

pO_2 room air lower than expected but not grossly hypoxic

Pulse Sats 99% > measured sats/expected for pO2, ie saturation gap
        Left shift of O_2-Hb curve
                CO toxicity: HbCO level 26%
                hyperbaric O_2 indicated

A-a gradient   = 150 – (1.25 x 58) - 82
                = 150 – 75 – 82
                = -7
        i.e. normal
        no V/Q mismatch exists

3. Electrolytes

Normonatraemia
Normokalaemia
        note K normal but low for ph expect higher
        expect 5 + 5 x 0.5 = 7.5mmol/L

106

likely full body K deficient

watch and replace k as ph corrected

Mild hypochloraemia

electrical equilibrium/renal excretion RAGMA

Normal BSL

not cause for decreased gcs

Normal renal indices

not cause for acidosis

Critical hyperlactataemia

Likely cause RAGMA:

Type A - tissue hypoxia : **CYANIDE** toxicity (note normoxic tissue hypoxia in setting of fire), also consider hypo-perfusion secondary shock/hypotension

Type B_1 sepsis/underlying disease (eg liver)

Type B_2 drug OD (salycilates, tox alcohols etc)

4. Interpret

This is a critically ill patient

respiratory acidosis

may need ventilatory support

RAGMA

likely lactic with N BSL and renal indices

carbon monoxide poisoning plus neurologic compromise

high flow O_2 indicated

critical lactic acidosis suggests cyanide toxicity in context of fire: empiric treatment with hydroxycobalamin and thiosulfate is indicated

Also consider other causes decreased GCS: closed head injury/trauma

A-a gradient in kPa

A-a grad

$= 21\ kPa - 1.25 \times 7.7 - 10.9$

$= 21 - 10 - 10.9$

$= 21 - 20.9$

$= 0.1$

Not elevated

107

COMMENTS

A 40 year old man is brought by ambulance to your emergency department after a <u>fire at a factory</u>. He has no airway compromise or evidence of cutaneous burns, but he is <u>GCS 12 and acutely confused</u>. His blood pressure is 120/80 and his oxygen saturation is 99%.

This is a somewhat obtuse question, but the mention of a fire should raise the spectre of carbon monoxide poisoning. This is clearly evident in the elevated HbCO level given in the question. However, astute candidates will note that there is a discrepancy between the observed oxygen saturation and the calculated oxygen saturation: the so called "saturation gap". There is also a substantial difference between the measured arterial oxygen tension and tissue perfusion, as measured by an extreme lactic acidosis. In other words there is a normoxic tissue hypoxia.

The saturation gap is due to a misread from the pulse oximeter, and strongly implies the presence of an abnormal form of haemoglobin, usually HbCO. It is strongly suggestive of carbon monoxide poisoning[46]. High flow oxygen is indicated as a minimum. The indications for hyperbaric oxygen remain controversial[47] but probably apply in this case.

The phenomenon of normoxic tissue hypoxia (the presence of adequate oxygen in the tissues but the inability to metabolize it) suggests cyanide toxicity. In developed countries cyanide poisoning is most commonly associated with fires and industrial accidents[48]. Cyanide uncouples oxidative phosphorylation and electron transport, and is therefore non responsive to conventional measures of improving oxygenation. It requires empiric treatment with one of several antidotes, most commonly hydroxocobalamin and sodium thiosulfate[31].

Strong candidates will discuss both these possibilities, as well as other differentials for the patients decreased level of consciousness, including head injury.

---

[46] Akhtar J, Johnston BD and Krenzelok EP. Mind the gap. *J Emerg Med* 2007; 33(2):131-2.
[47] Kao LW and Nanagas KA. Carbon monoxide poisoning. *Emerg Med Clin North Am* 2004; 22(4):985-1018
[48] Megarbane B, Delahaye A, Goldgran-Toledano D and Baud FJ. Antidotal treatment of cyanide poisoning. *J Chin Med Assoc* 2003; 66(4): 193-203

PROBLEM 25

A 35 year old male multi-trauma victim attends your hospital with the ambulance service after a high speed MVA. The ambulance have given him several liters of crystalloid en route to hospital. He has obvious chest injuries.

His vital signs are:

HR	110	/min
BP	95/75	mmHg
RR	34	/min
Sats	92%	40% O^2
T	35.4	°C
GCS	7 ($E_1V_2M_4$)	

Urgent arterial blood gas analysis is undertaken and the results are shown below.

Describe and interpret them.   (70%)

Outline your management priorities.   (30%)

FiO_2	0.4		
pH	7.09		
PO_2	76	mmHg (10.1	kPa)
PCO_2	65	mmHg (8.7	kPa)
HCO_3^-	14	mmol/L	
BE	-8		
O_2 Sats	92%		
Na^+	135	mmol/L	
K^+	4.2	mmol/L	
Cl^-	103	mmol/L	
Glucose	6.2	mmol/L	
Lactate	4.1	mmol/L	
Hb	82	g/dL	

109

ANSWER

1. Acid base status

Severe acidaemia (pH 7.09)

Severe hypercarbia
  Respiratory acidosis
  Expect HCO_3^- for CO_2 65
    Acute   = 24 + 2.5 x 1 = 29mmHg
    Chronic = 24 + 2.5 x 4 = 34mmHg

Moderate decrease in HCO_3^-
Metabolic acidosis
  Expect CO_2 = 8 + 1.5 x 14
        = 8 + 21
        = 29
  AG   = 135 – 14 – 103
        = 135 – 117
        = 18 → RAGMA
  DR   = (18-12)/(24-14)
        = 6/10
        = 0.6
        Implies concurrent NAGMA and RAGMA

Therefore diagnose triple disturbance:
    acute respiratory acidosis
    RAGMA
    NAGMA
    No compensation

2. Oxygenation

Significant hypoxaemia on supplemental O_2

A-a gradient   = 300 – 1.25 x 65 – 76
        = 300 – 80 – 76
        = 300 – 156
        = ~134

Expect A-a gradient 35yo male = 35/4 + 4 = 9 + 4 = 13
Significant V/Q mismatch
    Given Hx concern RE
        Pulmonary contusion/flail segment
        ARDS/SIRS
        Aspiration

110

3. Electrolytes

Normonatraemia
Normokalaemia
      Expect K^+ for pH = 5 + 2.5 x 0.5 = 6.25 mmol/L
      Hypokalaemic for pH : correct as pH corrected
Normochloraemia
Normoglycaemia
Severe hyperlactataemia
      Likely cause for RAGMA
Profound anaemia
      Probably due to significant haemorrhage
      Needs transfusion
            Be prepared for massive transfusion protocol

4. Interpret

Critically injured patient: hypotensive, hypoxic, tachycardic, hypothermic, obtunded
      Critical triad: hypothermic, acidotic → check coag status!

ABG shows
      Respiratory acidosis
      RAGMA
      NAGMA

Respiratory acidosis plus raised A-a gradient
      Pulmonary contusion
      Haemo/pneumothorax
      ARDS/APO
      Other causes – fat embolism, PE, resp membrane defect
      Poor prognosis if closed head injury

RAGMA
      Hyperlactataemia
            Hypovolaemic shock
            Poor prognostic indicator
NAGMA
      Excessive crystalloid administration
      Renal failure
      Other source HCO_3^- loss

Management priorities

Secure airway (GCS <8)
    Ventilate to reverse resp acidosis
    Decompress chest if required

Resuscitate: blood, massive transfusion protocol
    Warm fluid
    Tranexamic acid

Transfer to definitive care
    OT, if stabilizes consider CT, otherwise FAST
Social worker for family

A-a gradient in kPa

A-a grad
    = 40 kPa − 1.25 x 8.7 − 10.1
    = 40 − 10.5 − 10.1
    = 40 − 20.6
    = 19.4 kPa

Grossly elevated

COMMENTS

A 35 year old male multi-trauma victim attends your hospital with the ambulance service after a high speed MVA. The ambulance have given him several litres of crystalloid en route to hospital. He has obvious chest injuries.

His vital signs are:

HR	*110*
BP	*95/75*
RR	*34*
Sats	*92% 6L*
T	*35.4°C*
GCS	*7 ($E_1V_2M_4$)*

The stem provides a familiar scenario to experienced emergency physicians. Based on the vital signs and the cue regarding chest injuries significant disturbances of the A-a gradient should be anticipated. The association of the administration of saline with a normal anion gap acidosis has already been discussed, and hypoperfusion is an important cause of a hyperlactataemic RAGMA. Systematic working of the question will reveal all of these abnormalities without too much effort.

Strong candidates will mention some of the key issues in care of the trauma patient. The deleterious triad of acidosis, hypothermia and coagulopathy in a trauma patient is well described[49]. Key priorities will be management of the airway[50], re-warming and correction of the coagulopathy, administration of tranexamic acid[51], transfer to definitive care and care of the patient's family. As the stem only asks for an "outline" of these issues they can be dealt with using broad brush strokes and little detail.

---

[49] Rossaint R, Cerny V, Coats TJ, Duranteau J, Fernandez-Mondejar E, Gordini G et al. Key issues in advanced bleeding care in trauma. *Shock* 2006; 26:322-31
[50] Gentleman D, Deaden M, Midgelu S and Maclean D. Guidelines for resuscitation and transfer of patients with serious head injury. *BMJ* 1997; 307:547-552
[51] The CRASH-2 trial collaborators. Effects of tranexamic acid on death, vascular occlusive events, and blood transfusion in trauma patients with significant haemorrhage (CRASH-2): a randomized, placebo-controlled trial. *The Lancet* 2010; 376(9734): 23 - 32

THE APPENDICES

"Success is not final, failure is not fatal: it is the courage to continue that counts"

Winston Churchill

Appendix 1: anion gap and delta ratio

A note on Anion Gap: what are the correct numbers?[52]

Candidates often express a degree of confusion about what constitutes the "normal" value of anion gap, and whether it should be calculated using potassium or not.

The authors' personal opinions are that adding complexity to an already stressful situation (ie the fellowship exam) is unnecessary, and thus any calculations should be as simple as possible. It is surprising under time pressure how difficult adding 4 and 17 can be!

Therefore this book uses an anion gap of 12, based on a calculation without potassium:

$$\text{ANION GAP} = [Na+] - ([Cl^-] + [HCO_3-])$$

Reference ranges:

> 12: indicates a raised anion gap metabolic acidosis (RAGMA). A delta ratio should be calculated to evaluate the concurrence of either a NAGMA or a metabolic alkalosis

< or equal to 12: indicates a NAGMA. A delta ratio is not performed

It is worth noting however that a NAGMA can present with a mildly raised anion gap (above 12). This is because there is not a discrete anion gap cut off for "normal" or "raised" values. Most papers specify an anion gap without K^+ as having a normal value of 12 +/- 4. Thus even in the event of an anion gap greater than 12, the possibility of a NAGMA should be considered, and a delta ratio calculation performed to evaluate for the possibility.

For those more interested again in the subject a good place to start is reading the reference below.

---

[52] Lolekha PH, Vanavanan S and Lolekha S. (2001) Update on value of the anion gap in clinical diagnosis and laboratory evaluation. *Clin Chim Acta* 307 (1-2): 33 – 36

The Delta Ratio[53]

Delta ratio is defined as (change in AG)/(change in HCO_3^-). It is calculated by:

$$\frac{AG - 12}{24 - [HCO_3^-]}$$

A delta ratio is interpreted as follows:

< 0.4 (large drop in HCO_3^- without change in AG) = NAGMA

0.4 – 0.8 (intermittent drop in AG compared to HCO_3^-) = NAGMA + RAGMA

1.0 (equal drop in HCO_3^- and AG) = RAGMA

2.0⁺ (lower drop HCO_3^- in than expected for AG) = there is a previously existing or concurrent metabolic alkalosis occurring. Mathematically this ratio occurs *because the resting HCO_3^- for the patient is greater than 24mmol/L*

[53] Kim HY, Han JS, Jeon US, Joo KW, Earm JH, Ahn C et al. (2001). Clinical significance of the fractional excretion of anions in metabolic acidosis. *Clin Nephrol* 2001 55(6): 448-52

Appendix 2: assessing compensation

The reference[54] given is not the original paper on correcting acid base balances, but provides a very good overview of acid-base balance in general, and correction rules in particular. The correction calculations follow.

For primary metabolic acidosis:

$$\text{Expected } CO_2 \text{ in mmHg} = 8 + 1.5 \times [HCO_3^-]$$

For primary metabolic alkalosis:

$$\text{Expected } CO_2 \text{ in mmHg} = 21 + 0.7 \times [HCO_3^-]$$

For primary respiratory acidosis:

$[HCO_3^-]$ increases by 1mmol/L per 10mmHg CO_2 above 40mmHg

$[HCO_3^-]$ increases by 4mmol/L per 10mmHg CO_2 above 40mmHg

For primary respiratory alkalosis:

$[HCO_3^-]$ decreases by 2mmol/L per 10mmHg CO_2 below 40mmHg

$[HCO_3^-]$ decreases by 5mmol/L per 10mmHg CO_2 below 40mmHg

---

[54] Sood P, Gunchan P and Sandeep P (2010). Interpretation of arterial blood gas. *Indian J Crit Care Med* 14(2): 57 - 64

Appendix 3: notes on sundry calculations

A-a Gradient[55]

The equation to calculate an A-a gradient in mmHg is:

$$[(760\text{-vapour pressure}) \times FiO_2 - 1.25 \times PaCO_2] - PaO_2$$

One of the hardest things in the exam is trying to calculate the P_AO_2 ([760-vapour pressure] x FiO_2) for an FiO_2 that's not room air.

This is the most obvious instance in the examination where approximation is acceptable:

- If FiO_2 = 21% (room air) then the first number in the equation (760-VP x FiO_2) is 150mmHg.
- By dead reckoning, FiO_2 40% then the first number is doubled: 300mmHg
- Similarly on 28% O_2 (somewhere in the ballpark of 1.5 x 21…) 225mmHg (1.5 x 150mmHg) can be used

A standard comment in the event of an elevated A-a gradient is:

"A-a gradient elevated. .: V/Q mismatch:
- *APO*
- *ARDS*
- *PE*
- *LRTI*
- *Other respiratory membrane disease*

There is some discussion of this particularly as it pertains to pneumonia in the associated reference.

For those using the calculations in kPa it is important to realise that atmospheric pressure (after water vapour is removed) at sea level is 101kPa. This means that the first number in the A-a gradient

---

[55] Moammar MQ, Azam HM, Blamoun AI, Rashid AO, Ismail M, Khan MA et al. (2008). Alveolar-arterial oxygen gradient, pneumonia severity index and outcomes in patients hospitalized with community acquired pneumonia. *Clin Exp Pharmacol Physiol* 35(9): 1032-7

calculation (the FiO_2) is equal to the percentage of administered oxygen. There is no easily applied formal corrective formula for age when kPa are used to calculate A-a gradient, and normal values cited in the literature vary. As a rule of thumb young people should have an A-a gradient between 2.5 – 3.0 kPa, while older people (60+) should have an A-a gradient <4.0 kPa[56].

Electrolytes

Sodium[57]

$$\text{Corrected Na}^+ = \text{Na}^+_m + (\text{Glucose} - 5)/3.$$

This is very important for DKA cases. The experimental numbers for absolute sodium vary but the equation above is a useful approximation for the exam.

Potassium[58]

Potassium homeostasis is complex, and varies with the type and mechanism of acidaemia or alkalaemia. Practically, for exam purposes for each pH fall below 7.4, K^+ should rise by 0.5mmol above 5.0mmol/L. Therefore an appropriate comment would be along the lines of:

> K^+ 4.0mmol/L = normal.
> Expect K^+ for pH 7.2 = 5.0 + 2 x 0.5 = 6.0mmol/L.
> ∴ pt relatively hypokalaemic. Watch/replace K^+ as pH corrected.

Renal indices[59]

When assessing urea to creatinine ratio the examination answer is almost always ~100 and therefore suggests pre-renal failure. Again this is a gross simplification, but unless the number is very low (<50 indicating intrinsic renal failure) it is a useful mechanism for the exam.

---

[56] Williams AJ. Assessing and interpreting arterial blood gases and acid-base balance. *BMJ* 1998; 317(7167): 1213 – 1216.3

[57] Hillier TA, Abbott RD and Barrett EJ. (1999) Hyponatraemia: evaluating the correction factor for hyperglycaemia. *Am J Med* 106(4): 399 – 403.

[58] Rastegar A. Clinical Methods: the History, Physical and Laboratory Examinations. 3rd Ed. Chapter 195 Boston, *Butterworths*; 1990

[59] Feinfeld DA, Bargouthi H, Niaz Q, Carvounis CP. (2002) Massive and disproportionate elevation of blood urea nitrogen in acute azotaemia. *Int urol nephrol* 34(1) 143:145

Osmolality[60]

The only reason to give candidates a measured osmolality in the question stem is to enable calculation of an osmolar gap, which is a two step process:

$$\text{Calculated osmolality} = 2 \times Na^+ + urea + glucose$$

$$\text{osmolar gap} = OSM_m - OSM_c.$$

 If there is a measured osmolality in the stem you would be foolish not to perform these calculations.

---

[60] Hoffman RS, Smilkstein MJ, Howland MA and Goldfrank LR. (1993) Osmol gaps revisited: normal values and limitations. *Clin Toxicol,* 31: 81-93.

Appendix 4: ACEM terminology

"It is always important to read a question carefully, and to understand these terms in the context of that question. The Fellowship Examination Committee and the examiners are instructed to be rigorous in the use of these terms. Candidates are advised to do similarly."[61]

Assessment: History taking, physical examination and investigations.

Describe: State the characteristics or appearance of the subject, including relevant negatives.

Discuss: Examine the pros and cons of each of the alternatives asked for on a subject.

Disposition: Where the patient is sent following care in the emergency department, including follow - up if discharged.

Interpret: State a conclusion or conclusions which includes a differential diagnosis, but excludes management.

Investigations: Specific tests undertaken to make a diagnosis or monitor the patient's condition.

List: A numerical ordering of related items.

Management: Those aspects of care of the patient encompassing treatment, supportive care and disposition.

Outline: A brief description of the subject.

Protocol: A set of instructions on how to deal with a particular situation.

Treatment: Measures undertaken to cure or stabilize the patient's condition

---

[61] Australasian College for Emergency Medicine. Examination Handbook v02. Published online and retrieved 18 November 2013 from http://www.acem.org.au/media/publications/FEH02_v02_Fellowship_Exam_Structure__Mar-12_.pdf

Appendix 5: Luke Lawton's lists of causes of acid-base disturbances

The following lists are taken from Luke's study notes when preparing for the ACEM fellowship exam. They are illustrative of the type of list for causes of acid-base disorders that candidates should assemble in their minds prior to the exam. In no way are the lists definitive in content or scope and are not specific to any particular reference. The information is essentially an amalgamation of multiple sources compiled during study, arranged in an order pleasing to the author.

Causes of RAGMA

1. Lactic Acid
>A: cyanide toxicity, arterial occlusion (mocontorio ischaemia), hypovolaemia, profound anaemia
>
>B_1: liver failure (paracetamol poisoning), sepsis
>
>B_2: most drugs (methanol, ethylene glycol, paraldehyde, paraquat, iron, isoniazid, metformin)
>
>B_3: inborn errors of metabolism

2. Ketoacids
>Diabetic ketoacidosis
>Starvation/alcoholic ketoacidosis

3. Salycilates

4. Uraemic renal failure

Causes of NAGMA

1. HCO_3^- loss (most common)
      Excessive diarrhoea
      Renal tubular acidosis
      Fistulas (pancreatico-duodenal, ureto-enteric)

2. Excessive 0.9% NaCl administration

3. Endocrinopathies
      Mineralocorticoid or glucocorticoid deficiency
      Adrenal crisis

4. Drugs
      Carbonic anhydrase inhibitors (acetazolamide)
      Spironolactone

Causes of metabolic alkalosis

1. HCl loss (most common)
      Vomiting

2. Drug induced
      Diuretics (esp frusemide)
      Laxative abuse
      Milk-alkali syndrome
      NaHCO3 therapy

3. Endocrinopathies
      Mineralocorticoid or glucocorticoid excess
      Bartter's syndrome

Causes of respiratory alkalosis

1. Central stimulation of respiratory drive
   Head injury/ICH
   Anxiety/pain
   Salicylate intoxication
   Sympathomimmetic intoxication (MDMA etc)

2. Profound hypoxia causing peripheral respiratory stimulation
   PE, pneumonia, pulmonary oedema, acute asthma, ARDS

Causes of respiratory acidosis

1. Acute on chronic exacerbation of COAD

2. Airway disorders
   Acute asthma
   Airway obstruction

3. Depression of central respiratory drive
   Drugs (opiates, sedatives)
   ICH, CNS injury

4. Lung insults
   Trauma (pulmonary contusion, pneumothorax, haemothorax, flail segment)
   Lung membrane diseases (pulmonary oedema, lower respiratory tract infection, ARDS, aspiration)

5. Inadequate respiratory effort
   Toxins (organophospates, elapid envenomation)
   Myopathies, Guillain-Barre syndrome

Appendix 6: examination reference values

It is important to have a set of easily remembered reference values that can be readily applied during the exam. The following have been used in the preparation of answers in this book.

pH	7.35 – 7.45	
PO$_2$	100	mmHg
PCO$_2$	35 – 45	mmHg
HCO$_3^-$	24	mmol/L
Base Excess	-3 to + 3	
Lactate	<2	mmol/L
Na$^+$	135 - 145	mmol/L
K$^+$	3.5 – 4.5	mmol/L
Cl$^-$	100 – 115	mmol/L
Creatinine	<100	mmol/L
Urea	<7	mmol/L
Glucose	<6.5	mmol/L

About the authors

Dr Luke Lawton holds degrees in biochemistry, medicine, surgery and public health, and is a Fellow of the Australasian College for emergency medicine. He is an active emergency physician at Redcliffe Hospital in northern Brisbane, as well as a retrieval physician with Careflight Medical Services. His educational interests include teaching at undergraduate level through the university, as well as preparation of ACEM fellowship candidates through his local health service and the resus.com.au fellowship course. Clinically his interests are the practice of intelligent medicine, and doing the right thing for his patients.

Dr Corinne Ryan is a Fellow of the Royal Australian College of Physicians and specialises in medical oncology. She has a biochemistry degree from Queensland University of Technology, and studied medicine at James Cook University. She is enthusiastic about clinical teaching of undergraduates and doctors in training, and has provided specialist medical comment on many of the topics in this book.

www.ingramcontent.com/pod-product-compliance
Lightning Source LLC
Chambersburg PA
CBHW021602210326
41599CB00010B/556